IX International Symposium "Modern Problems of Laser Physics" (MPLP 2021)

Journal of Physics: Conference Series Volume 2067

Novosibirsk, Russia
22 – 29 August 2021

ISBN: 978-1-7138-5038-0
ISSN: 1742-6588

Printed from e-media with permission by:

Curran Associates, Inc.
57 Morehouse Lane
Red Hook, NY 12571

Some format issues inherent in the e-media version may also appear in this print version.

Institute of Physics
Dirac House, Temple Back
Bristol BS1 6BE UK

Phone: 44 1 17 929 7481
Fax: 44 1 17 920 0979

techtracking@iop.org

Additional copies of this publication are available from:

Curran Associates, Inc.
57 Morehouse Lane
Red Hook, NY 12571 USA
Phone: 845-758-0400
Fax: 845-758-2633
Email: curran@proceedings.com
Web: www.proceedings.com

TABLE OF CONTENTS

Journal of Physics: Conference Series **2067** (2021) 011001 doi:10.1088/1742-6596/2067/1/011001

Preface

General Information

This volume of the Journal is devoted to the IX International Symposium and Young Scientists School “Modern Problems of Laser Physics” (MPLP-2021)[1] that was held in Novosibirsk, Russia, 22-28 August 2021.

The MPLP symposium has a long history starting from 1995 and gathers scientists from many courtiers who carrying out their researches in the forefront of laser physics, quantum metrology and high-resolution spectroscopy, physics of ultracold atoms, molecules and ions, atom optics, ultrafast phenomena and attoscience, quantum optics and information, nonlinear optics and applications of laser radiation from THz to UV radiation ranges in medicine, geophysics, chemistry and microbiology. This year the sessions devoted to advances in laboratory space plasma physics with lasers was also included in the list of topics of MPLP symposium.

The Scientific Program of MPLP-2021 includes invited and oral talks selected from contributed papers, as well as several Poster Sessions. A Scientific School for Young Scientists was also take place during the Symposium. Following the tradition of MPLP Symposia, the scientific program comprises single sessions of invited and oral presentations for each topic, running sequentially. This year online access to MPLP was provided for participants who can’t attend the symposium in-person and many of the invited talks were done online.

Organizers:

The Symposium was traditionally organized by the Institute of Laser Physics of the Siberian Branch of Russian Academy of Sciences[2] as a host organization in close

- *Institute of Spectroscopy, RAS, Troitsk, Moscow, Russia*[3]
- *Federal State Unitary Enterprise "VNIIFTRI", Mendeleevo, Moscow region, Russia*[4]
- *Novosibirsk State University, Novosibirsk, Russia*[5]

[1] mplp2021.laser.nsc.ru
[2] www.laser.nsc.ru
[3] isan.troitsk.ru
[4] www.vniiftri.ru
[5] www.nsu.ru

Journal of Physics: Conference Series **2067** (2021) 011001 doi:10.1088/1742-6596/2067/1/011001

Committees

1. International Advisory Committee

The Organizers thank all members of the International Advisory Committee for their valuable contributions to the Program of the meeting.

Dmitry Budker	*Helmholtz Institute Mainz, Johannes Gutenberg University, Germany*
Ekkehard Peik	*Physikalisch-Technische Bundesanstalt (PTB), Braunschweig, Germany*
Victor Balykin	*Institute for Spectroscopy, Russia*
Alexey Zheltikov	*Lomonosov Moscow State University, Russia*
Leong Chuan Kwek	*Centre for Quantum Technologies, National University of Singapore*
Ken-ichi Ueda	*Institute for Laser Science, University of Electro Communications, Tokyo, Japan*
Victor Zadkov	*Institute of Spectroscopy RAS, Troitsk, Russia*
Masahiko Tani	*Research Center for Development of Far-Infrared Region, University of Fukui*
John Kitching	*Time and Frequency Division, NIST, Boulder, USA*
Igor Ryabtsev	*Rzhanov Institute of Semiconductor Physics SB RAS, Russia*
Mikhail Starodubtsev	*Institute of Applied Physics of the Russian Academy of Sciences*
Nikolai Kolachevsky	*P.N. Lebedev Physical Institute RAS, Moscow, Russia*
Gerhard Leuchs	*Planck Institute for the Science of Light, Erlangen-Nuremberg, Germany*
Martial Ducloy	*Laboratoire de Physique des Lasers, Universite Paris-Nord, France*
Andrey Turlapov	*Institute of Applied Physics of the Russian Academy of Sciences, Nizhny Novgorod, Russia*

2. Local Organizers

Chairman of the Symposium:

Prof. Sergei N. Bagayev
Academician
Scientific Director of the Institute of Laser Physics SB RAS

Organizing Committee:

Prof. Alexey V. Taichenachev
Chairman of the Local Program Committee

Corresponding Member of the Russian Academy of Sciences, Director of the Institute of Laser Physics SB RAS

Prof. Ildar Shaikhislamov
Vice Chairman of the Local Program Committee

Deputy Director for Research of the Institute of Laser Physics

Assoc. Prof., Oleg Prudnikov
Scientific Secretary of the Symposium

Principal Researcher of the Institute of Laser Physics

Dr. Roman Ilenkov
Scientific Secretary of the School for Young Scientists

Researcher of the Institute of Laser Physics

Members of the Local Program Committee

Invited and Oral Speakers

Name	Organization	Title of the talk
Hidetoshi Katori	*Department of Applied Physics, Graduate School of Engineering, The University of Tokyo, Bunkyo-ku, Tokyo, Japan*	Transportable Optical Lattice Clocks to Test and Use Gravitational Redshift
Gerhard Leuchs	*Max Planck Institute for the Science of Light, Erlangen-Nuremberg, Germany*	Interfering the bright and the dark – a new route to implementing Kerr squeezing
Ekkehard Peik	*Physikalisch-Technische Bundesanstalt, Braunschweig, Germany*	Improved Limits for Violations of Local Position Invariance from Atomic Clocks
Rodolphe Boudot	*FEMTO-ST, CNRS, UBFC, ENSMM, Besançon cedex, France*	Cs microcell microwave and optical frequency references at FEMTO-ST
Nils Huntemann	*Physikalisch-Technische Bundesanstalt, Braunschweig, Germany*	Optical clocks with trapped Yb^{+} ions
John Kitching	*Time and Frequency Division, NIST, Boulder, USA*	Compact, two-photon optical clock
Songbai Kang	*Innovation Academy for Precision Measurement Science and Technology, Wuhan, China*	In-progress report of miniaturized vapor-cell Rubidium optical clock at APM
Sergey Chepurov	*Institute of Laser Physics SB RAS, Novosibirsk, Russia*	Development of an optical frequency standard with a single 171Yb+ ion
Hua Guan	*Innovation Academy for Precision Measurement Science and Technology, Chinese Academy of Sciences, Wuhan, China*	The progress of Ca+ optical clocks
Valeri Yudin	*Institute of Laser Physics SB RAS, Novosibirsk, Russia*	Autobalanced Ramsey Spectrscopy
Vitaly Palchikov	*National Research Institute for Physical-Technical and Radiotechnical Measurements, Mendeleevo, Russia*	Current state and development of the State primary standard of unit of time, frequency and national time scale
Takashi Mukaiyama	*Graduate School of Engineering Science, Osaka University, Osaka, Japan*	Three-dimensional matter-wave interferometry of a trapped single ion
Lin Zhou	*Wuhan Institute of Physics and Mathematics, Innovation Academy for Precision Measurement Science and Technology, Chinese Academy of Sciences, Wuhan, China*	United test of the equivalence principle using mass and internal energy specified atoms
Ernst Rasel	*Institute of Quantum Optics, Hannover, Germany*	Twin-lattice interferometry
Zanon-Willet	*Sorbonne Université, Observatoire de Paris, Université PSL, CNRS, LERMA, Paris, France*	Hyper Ramsey-Bordé matter-wave interferometry
Chang Hee Nam	*Center for Relativistic Laser Science, Institute for Basic Science, Korea*	Ultrahigh intensity multi-PW laser for strong field physics research
Nikolai Rozanov	*Ioffe Institute, St. Petersburg, Russia;*	From few-cycle to subcycle and unipolar radiation pulses
Georg Becker	*Institut für Optik und Quantenelektronik, Friedrich-Schiller-Universität Jena, Jena, Germany*	Laser-Driven Proton Acceleration with Water Droplets and Thin Foils
Vladimir Trunov	*Institute of Laser Physics SB RAS, Novosibirsk, Russia*	Limitations in the implementation of coherent pulse combining with peak power up to multi-petawatt level
Stefan Wabnitz	*Sapienza University of Rome, Rome, Italy*	Multidimensional laser beam shaping with multimode optical fibers

Nobuyuki Matsumoto	*Dept. Physics, Gakshuin University, Tokyo, Japan*	High Q mg-scale monolithic pendulum for quantum-limited sensing
Denis Brazhnikov	*Institute of Laser Physics SB RAS, Novosibirsk, Russia*	High-contrast zero-field level-crossing resonances in cesium vapors for atomic magnetometry applications
Anton Vershovskii	*Ioffe Institute, St. Petersburg, Russia*	Optically pumped magnetic field sensors for magnetoencephalography and ultra-low field tomography
Andrey Goncharov	*Institute of Laser Physics SB RAS, Novosibirsk, Russia*	Cold atom interferometry: towards Mg optical standard and Rb quantum sensors of inertial forces
Oleg Prudnikov	*Institute of Laser Physics SB RAS, Novosibirsk, Russia*	Fast and deep laser cooling of ^{172}Yb ion in a radiofrequency trap
Anton Afanasiev	*Institute of Spectroscopy Russian Academy of Sciences, Moscow, Troitsk, Russia*	Atom femto trap: experimental realization and its spectroscopic perspectives
Xaioding He	*Wuhan Institute of Physics and Mathematics, APM, Chinese Academy of Sciences, Wuhan, China*	Coherently forming a single molecule in an optical trap
Hélène Perrin	*Laboratoire de physique des lasers, CNRS and Sorbonne Paris Nord University, Villetaneuse, France*	Physics in a bubble: from supersonic rotation to the effects of dimensional reduction
Tilman Pfau	*Physikalisches Institut and Center for Integrated Quantum Science and Technology, Universität Stuttgart, Germany*	A dipolar supersolid and a novel microscope to probe quantum gases
Leong Chuan Kwek	*Centre for Quantum Technologies, National University of Singapore, Singapore*	Chip-Based Quantum Cryptography
Igor Ryabtsev	*Rzhanov Institute of Semiconductor Physics SB RAS, Novosibirsk, Russia*	Towards quantum gates with single rubidium Rydberg atoms in an array of optical dipole traps
Ennio Arimondo	*Pisa University, Pisa, Italy*	Double dressing quantum control
Mark Saffman	*Department of Physics, University of Wisconsin-Madison, Madison, WI, USA*	Circuit model quantum computing with atomic qubits
Denis Sutyrin	*National Research Institute for Physical-Technical and Radiotechnical Measurements, Mendeleevo, Russia*	Towards optical time scale at VNIIFTRI
Dmitry Tregubov	*Lebedev Physical Institute, Moscow, Russia*	Search for magic wavelength near 1064 nm for thulium optical clock
Stepan Ignatovich	*Institute of Laser Physics SB RAS, Novosibirsk, Russia*	Intensity shifts in CPT Resonances on the ^{87}Rb D1 line with excitation by VCSEL modulated at frequencies 3.4GHz and 6.8GHz
Maxim Okhapkin	*Physikalisch-Technische Bundesanstalt, Braunschweig, Germany*	Towards direct laser excitation of the low-energy nuclear transition in ^{229}Th.
Vitaly Palchikov	*National Research Institute for Physical-Technical and Radiotechnical Measurements, Mendeleevo, Russia*	Towards a new definition of the second in the SI based on optical atomic clocks
Nataliya Rubtsova	*Rzhanov Institute of Semiconductor Physics SB RAS, Novosibirsk, Russia*	Semiconductor quantum well based shutters for NIR laser mode-locking with ~GHz repetition rate
Victor Zadkov	*Institute of Spectroscopy of the Russian Academy of Sciences, Troitsk, Moscow, Russia*	Quantum Optics of Quantum Emitters in Nanostructures
Yuri Kivshar	*Nonlinear Physics Center, Australian*	Metaphotonics and metasurfaces

	National University, Canberra, Australia	
Oleg Prudnikov	*Institute of Laser Physics SB RAS, Novosibirsk, Russia*	Guided-mode waves structure of electric and magnetic dipole resonances in metamaterials slab
Hiroaki Minamide	*RIKEN Center for Advanced Photonics, RIKEN, Aoba-ku, Sendai, Japan*	Backward optical parametric oscillator to generate tunable terahertz waves
Masahiko Tani	*Research Center for Development of Far-Infrared Region, University of Fukui, Fukui, Japan*	Terahertz emission from spintronic Fe/Pt bilayers
Alexey Balakin	*Lomonosov Moscow State University, Moscow, Russia*	Nonlinear optical phenemena with terahertz pulses
Olga Cherkasova	*Institute of Laser Physics SB RAS, Novosibirsk, Russia*	The Interaction of THz radiation with molecules and cells
Feng Song	*Nankai University, Tianjin, China*	Flexible color-tunable composite for broadband response and diverse application
Dmitry Kolker	*Institute of Laser Physics SB RAS, Novosibirsk, Russia*	Tunable MID-IR laser sources for trace-gas analysis
Nadezhda Kostyukova	*Institute of Laser Physics SB RAS, Novosibirsk, Russia*	Near-degenerate narrowband tunable MgO:PPLN OPO generating radiation with wavelength of 2.1 μm
Ildar Shaikhislamov	*Institute of Laser Physics SB RAS, Novosibirsk, Russia*	Laser experiments to model plasma releases in the Earth magnetosphere: past, present and unsolved problems
Julien Fuchs	*LULI - CNRS, CEA, UPMC Univ Paris 06: Sorbonne Université, Ecole Polytechnique, Institut Polytechnique de Paris, France*	Laboratory evidence for proton energization by collisionless shock surfing
Mikhail Starodubtsev	*Institute of Applied Physics, Russian Academy of Sciences, Nizhny Novgorod, Russia*	Laboratory modeling of protostellar jet formation in diverging poloidal magnetic field
Konstantin Burdonov	*LULI - CNRS, CEA, UPMC Univ Paris 06: Sorbonne Université, Ecole Polytechnique, Institut Polytechnique de Paris, France*	Laboratory modelling of matter accretion using laser-driven plasma coupled with a strong magnetic field source
Dmitriy Bisikalo	*Institute of Astromony RAS, Moscow, Russia*	Gasdynamics of plasma envelopes of hot exoplanets
Andrey Divin	*St. Petersburg State University, Physics Department, St. Petersburg, Russia*	Three-dimensional Particle-in-Cell (PIC) simulations of minimagnetosphere formation in the KI-1 experiment
Andrey Sladkov	*Institute of Applied Physics RAS, Russia*	Three-dimensional numerical modeling of magnetic reconnection in laser-induced high energy density plasmas
Vojtech Horny	*LULI-CNRS, CEA, UPMC Univ Paris, France*	Investigation of the prospects for achieving nucleosynthesis of heavy elements in the laboratory
Frederico Fiuza	*SLAC National Accelerator Laboratory, California, USA*	Electron acceleration in laboratory-produced turbulent collisionless shocks
Deepak Kumar	*Johns Hopkins University, USA*	Radiative characterization of a laser driven shock
Derek Schaeffer	*Princeton University, Princeton, New Jersey, USA*	Laboratory Studies of Laser-Driven, High-Mach-Number Magnetized Collisionless Shocks

School for Young Scientists

The School for Young Scientists was organized as a part of the Symposium. Several top-ranking scientists delivered their lectures on "hot" topics of modern laser physics and applications:

Prof. Masahiko Tani

Research Center for Development of Far-Infrared Region
University of Fukui, Japan

Lecture: "Nonlinear Optics for THz wave generation"

Prof. ***Dmitry Budker***

Helmholtz Institute, Johannes Gutenberg University, Mainz

Lecture: "In search of ultralight bosonic dark matter"

Prof. Gerd Leuchs

Max Planck Institute for the Science of Light, Erlangen-Nuremberg, Germany

Lecture: "Oscillations in space and time through interference"

The young scientists also give them contribution by presenting their works by oral talk during the school sessions.

Results

About 130 participants took part at the MPLP-2021 from 10 countries: China, France, Germany, Italy, Japan, Australia, Russia, Singapore, South Korea, and USA. The Symposium Program was consisted of invited (30 min), oral (15 min) and poster presentations.

The Organizers would like to thank participants of the MPLP-2021 for their valuable contributions to success of the Symposium and their interest. We also appreciate the help of all persons, who were involved into the organizing process.

We are looking forward to future fruitful and interesting MPLP meetings!

The Organizers

Peer review declaration

All papers published in this volume of Journal of Physics: Conference Series have been peer reviewed through processes administered by the Editors. Reviews were conducted by expert referees to the professional and scientific standards expected of a proceedings journal published by IOP Publishing.

- **Type of peer review: Single-blind / Double-blind / Triple-blind / Open / Other (please describe)**

we used the ***Single-anonymous*** *peer review: authors' identities are known to the reviewers, reviewers' identities are hidden from authors with the following criteria for accepting/declining the manuscripts:*

<table>
<tr><td></td><td colspan="4"></td><td>“yes”</td><td>“no”</td></tr>
<tr><td>1</td><td colspan="4">Does this manuscript correspond to the subject of the MPLP symposium?</td><td>[ ]</td><td>[ ]</td></tr>
<tr><td>2.</td><td colspan="6">Assessment of the scientific level of work</td></tr>
<tr><td>2.1.</td><td colspan="4">Does the manuscript contain original results?</td><td>[ ]</td><td>[ ]</td></tr>
<tr><td>2.2.</td><td colspan="6">Indicate what these results are:</td></tr>
<tr><td>2.3.</td><td colspan="6">If the results are not original, indicate in which published works they are contained:</td></tr>
<tr><td>2.4.</td><td colspan="4">Are there any false statements in the article?</td><td>[ ]</td><td>[ ]</td></tr>
<tr><td>2.5.1</td><td colspan="4">Does this article convince the reviewer of the reliability of the results?</td><td>[ ]</td><td>[ ]</td></tr>
<tr><td>2.5.2</td><td colspan="4">Does this article convince the reviewer of the significance of the results obtained, sufficient for publication?</td><td>[ ]</td><td>[ ]</td></tr>
<tr><td>2.5.3</td><td colspan="4">Does this article convince the reviewer of the usefulness and relevance of the research?</td><td>[ ]</td><td>[ ]</td></tr>
<tr><td>2.6.</td><td colspan="4">Are there any references to the works of other authors, allowing you to correctly determine the place of this article among other publications?</td><td>[ ]</td><td>[ ]</td></tr>
<tr><td>2.7.</td><td>Your assessment of the scientific level of the manuscript as a whole (on a five-point system) (circle the required)</td><td>1</td><td>2</td><td>3</td><td>4</td><td>5</td></tr>
<tr><td>3.</td><td colspan="6">Assessment of the quality of the manuscript in accordance with the requirements of the JPCS</td></tr>
<tr><td>3.1.</td><td colspan="4">Does the title correctly reflect the main result of the work?</td><td>[ ]</td><td>[ ]</td></tr>
<tr><td>3.2.</td><td colspan="4">Does the Introduction state the purpose of the work?</td><td>[ ]</td><td>[ ]</td></tr>
</table>

Journal of Physics: Conference Series **2067** (2021) 011002 doi:10.1088/1742-6596/2067/1/011002

3.3.	Have the main results of the work been formulated in the Conclusion?				[ ]	[ ]
3.4.	Are corrections required					
	in the text?				[ ]	[ ]
	in formulas?				[ ]	[ ]
	in the figures?				[ ]	[ ]
	in the tables?				[ ]	[ ]
	in the list of references?				[ ]	[ ]
3.7.	Your assessment of the quality of the text of this article (on a five-point system) (circle the required)	**1**	**2**	**3**	**4**	**5**
4.	**Conclusion on the manuscript (check only one of points 4.1 - 4.4)**					
4.1.	The manuscript should not be published					[ ]
4.2.	The manuscript should be radically revised and the issue of its publication should be considered after a new review.					[ ]
4.3.	The manuscript can be published after revision of the text					[ ]
4.4.	The manuscript can be published without significant changes					[ ]
5.	**Referee comments: justification of negative ratings, detailed answers to questions, general opinion about the work, etc.**					

- **Conference submission management system:**
  the information on MPLP proceedings was published on the website of MPLP
  http://mplp2021.laser.nsc.ru/
  The manuscripts was accepted for review trough the symposium e-mail
  mplp2021@gmail.com

- **Number of submissions received: 22**

- **Number of submissions sent for review: 22**

- **Number of submissions accepted: 22**

- **Acceptance Rate (Number of Submissions Accepted / Number of Submissions Received X 100): 100**

- **Average number of reviews per paper: 2.4**

- **Total number of reviewers involved: 9**
- **Any additional info on review process:**

The Peer Review has been carried out by the Program Committee of the MPLP'21 symposium and reviewed in accordance with the Peer Review Policy **of papers published in the IOP Conference Series titles**

- **Contact person for queries:**
Name : Oleg Prudnikov
Affiliation: Institute of Laser Physics SB RAS, 15B Lavrentjeva Ave., Novosibirsk, Russia
Email : mplp2021@gmail.com

Specifics of using multifrequency pumping to register CPT resonances

K N Savinov[1], A K Dmitriev[1,2] and A V Krivetskii[3]

[1] Department of Laser Physics, Physical Engineering Faculty, Novosibirsk State Technical University, Novosibirsk, Russia
[2] Institute of Laser Physics SB RAS, Novosibirsk, Russia
[3] Department of Design and Technology of Radio Electronic Devices, Radio Engineering and Electronics Faculty, Novosibirsk State Technical University, Novosibirsk, Russia

k.n.savinov@yandex.ru

Abstract. When pumped by a diode laser, the injection current of which is modulated simultaneously by microwave and HF signals, CPT resonances are recorded in absorption, spaced from each other by an amount equal to half of the HF modulation frequency.

Keywords: CPT resonance, diode laser, frequency modulation.

1. Introduction

Since the creation of the first masers and lasers, continuous work has been carried out to improve the stability of quantum frequency standards [1]. Currently, rubidium clocks based on coherent population trapping (CPT) resonances are widely used [2, 3].

One of the main factors affecting the stability of frequency standards is light shift. The paper [4] describes a method for decreasing light shifts using multifrequency radiation of a femtosecond laser. In this case, the width of the emission spectrum significantly exceeds the value of the interval between optical transitions used for pumping CPT resonances. Also, most of the spectrum does not contribute to the pumping, which leads to a significant decrease in the signal-to-noise ratio. Moreover, other optical lines may be contained in wide spectrum of the femtosecond laser. Apparently, these facts are associated with the lack of successful experiments in this direction.

The spectrum matching condition is fulfilled in a diode laser with an external cavity, in which a multifrequency spectrum was realized with HF modulation of the injection current [6]. However, with this type of modulation, frequency adjustment is possible only by changing the cavity length, which is problematic [7]. It is quite simple to control the emission spectrum of a diode laser under the combined action of microwave and HF modulation of the injection current [8].

This paper presents the results of experiments on the registration of CPT resonances in the absorption in a cell with rubidium-87 under multifrequency pumping by a diode laser with an external cavity.

Journal of Physics: Conference Series **2067** (2021) 012001 doi:10.1088/1742-6596/2067/1/012001

2. Experimental setup

We used a diode laser with an external cavity represented by a diffraction grating, the length of which was set in such a way that the intermode interval was close to half the frequency of the clock transition in rubidium (3.4 GHz). Schematic of the experimental setup is shown in figure 1.

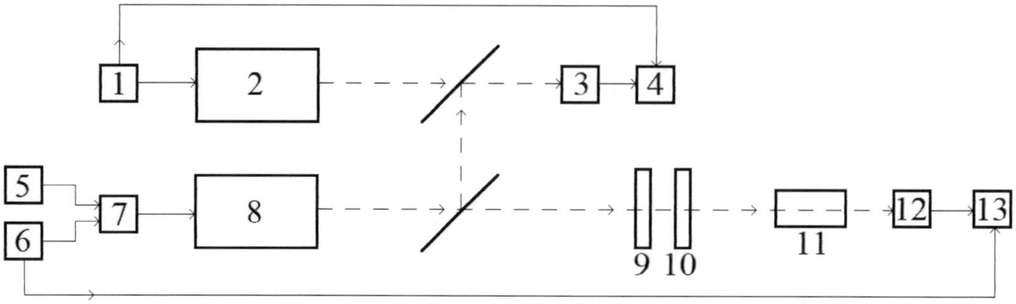

Figure 1. Schematic of the experimental setup. 1 - sawtooth voltage generator, 2 - heterodyne diode laser, 3, 12 - photodetectors, 4, 13 - digital oscilloscopes, 5 - HF signal source, 6 - microwave signal source, 7 - optical mixer, 8 - diode laser under study, 9 - polarizer, 10 - quarter-wave plate, 11 - cell with rubidium vapor. Dashed lines show optical connections, solid - electrical.

HF (5) and microwave (6) modulation signals through an optical mixer (7) were fed to a diode laser with an external cavity (8). To eliminate feedback, a polarizer (9) and a quarter-wave plate (10) are installed in the path of the beam. Then, the radiation entered cell (11) with Rb-85 and Rb-87 vapors with the addition of 15 Torr Ne as a buffer gas. The transmitted radiation was recorded by a photodetector (12) connected to a digital oscilloscope (13). The scanning signal of the microwave modulation frequency was also fed to the oscilloscope for synchronization.

To record the fine structure of the spectrum, a heterodyne diode laser (2) was used, the length of the external cavity of which was scanned using a sawtooth voltage generator (1). The beats were recorded by a photodetector (3) connected to a digital oscilloscope (4). The signal of the sawtooth voltage generator was used for synchronization.

3. Experimental results and discussion

The emission spectrum of diode laser under the combined action of microwave (3.417 GHz, 16 dBm) and HF (17 MHz, 1 dBm) modulation is shown in figure 2.

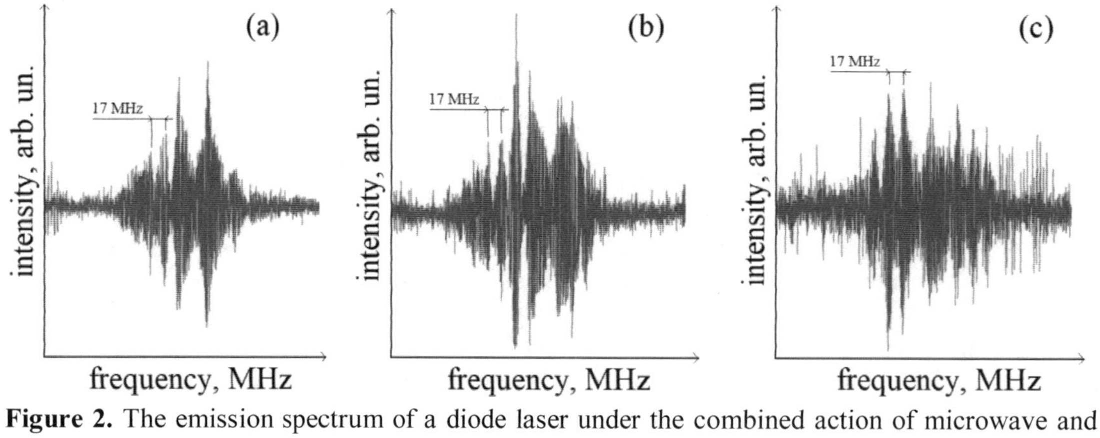

Figure 2. The emission spectrum of a diode laser under the combined action of microwave and HF modulation on: a - LF, b - carrier, c - HF bands.

It can be seen that at a modulation power of 1 dBm, each of the bands contains 5-8 HF components. A further increase in the HF modulation power leads to spectrum smearing. It should be noted that in each of the spectral bands, one of the HF components is absent in the central part due to the presence of amplitude modulation [9]. The resolution of HF components is limited by the bandwidth of the photodetector (2 MHz).

CPT resonances arise when the frequency difference between the interacting components is equal to the clock transition frequency. It is easy to show that the interval between CPT resonances is equal to half of HF modulation frequency [4].

When modulated radiation was passed through a cell with rubidium-87 vapor and a buffer gas, the structure of CPT resonances was detected, spaced from each other by half the HF modulation frequency (8.5 MHz). To record CPT resonances, the microwave modulation frequency was scanned with amplitude of 20 MHz, which leads to a change in the radiation power and the formation of a "pad" in the graph (figure 3).

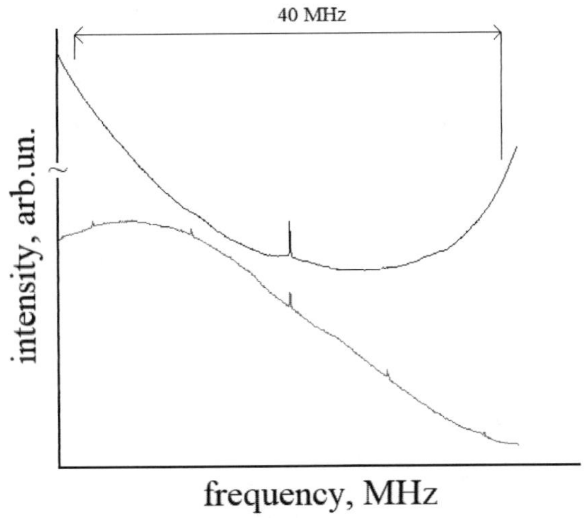

Figure 3. CPT resonances under microwave (black) and combined action of microwave and HF modulation (grey).

For clarity, a gap is added along the ordinate axis. A decrease in the signal when using combined modulation can be associated with the influence of RF modulation on the optical frequency of radiation [7].

A well-known technique for eliminating slope is to construct a differential signal. In our case, a record from a digital oscilloscope is a set of pointwise values, which means that a differential signal can be plotted as an algebraic difference between adjacent points of the graph (figure 4).

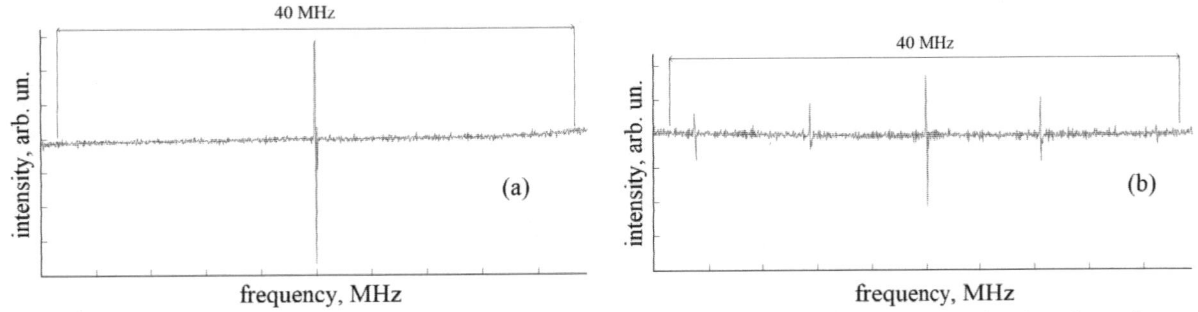

Figure 4. Derivative of CPT resonances with microwave modulation (a) and combined action of microwave and HF modulation (b).

With the addition of RF modulation, the amplitude of the central resonance decreases, since the dependence of the amplitude of the resonance on the intensity of the interfering fields has a threshold character [10]. However, this can lead to a decrease in the light shift, since the total amount will include shifts from each of the resonances, some of which will have a negative value. The amplitudes of the side resonances are significantly less than the amplitude of the central one, which may be due to the fact that, with a limited spectrum, the contribution to the side resonances is made by a smaller number of pairs of components.

4. Conclusion

When pumped by a diode laser, the injection current of which is modulated simultaneously by microwave and HF signals, CPT resonances are recorded in the absorption, spaced from each other by an amount equal to half the HF modulation frequency. An increase of the resonance amplitude is possible due to an increase in the radiation power; however, for a diode laser with an external cavity, this leads to a shift in the optical frequency.

Acknowledgement

This work was supported by the Ministry of Science and Higher Education of the Russian Federation, project no. FSUN-2020-0007.

References

[1] Quinn T 2005 *Metrologia* **42** 1
[2] Arimondo E and Orriols G 1976 *Lettere Al Nuovo Cimento* **17** 333
[3] Alzetta G, Gozzini A, Moi M and Orriols G. 1976 *Il Nuovo Cimento* **36** 5
[4] Baklanov E V, Bagaev S N, Dmitriev A K, Taichenachev A V and Yudin V I 2014 *Laser Physics* **24** 074007
[5] Vassiliev V V, Zibrov S A and Velichansky V L 2006 *Rev. Sci. Instrum.* **77** 013102
[6] Isakova A A, Golovin N N, Savinov K N and Dmitriev A K 2019 *Quant. Electon.* **49** 600
[7] Savinov K N, Dmitriev A K and Rundau A A 2021 *Optics and Spectroscopy* **129** 848
[8] Isakova A A, Savinov K N, Golovin N N, Sabakar K M, Dmitriev A K and Rundau A A 2020 *Rus. Phys. Jour.* **63** 171
[9] Kobayashi S, Yamamoto Y, Ito M and Kimura T. 1982 *IEEE J. Quant. Electron.* **18** 582
[10] Agap’ev B D, Gornyĭ M B, Matisov B G and Rozhdestvenskiĭ V Yu 1993 *Phys.-Usp.* **36** 763

About calculation of the ground-state energy of the helium atom

E V Baklanov[1,2], **P V Pokasov**[1] **and A V Taichenachev**[1,2]

[1]Institute of Laser Physics SB RAS, Novosibirsk 630090, Russia
[2] Novosibirsk State University, *Novosibirsk* 630090, Russia

E-mail: baklanov.ev@gmail.com

Abstract. Two versions of the numerical calculation of the ground state energy of the helium atom are compared. First, the nonrelativistic Schrödinger equation with a fixed nucleus is solved, and then the perturbation theory is used. Another version solves this problem exactly. Comparison shows that the difference between the calculation results is 94 kHz.

1. Introduction

Precision spectroscopy of the helium atom plays an important role in the theory of three-particle problems. The energy levels of such a system are expressed as the sum of nonrelativistic energy, relativistic corrections, and radiative corrections that include the Lamb shift. The accuracy with which the energy levels of the nonrelativistic problem are found determines the volume of all subsequent calculations, that is, the corrections that must be taken into account. In this work, two versions of the numerical calculation of the ground state energy of the helium atom are compared. One traditional one, where first the nonrelativistic Schrödinger equation with a fixed nucleus is solved, and then, to take into account its motion, the perturbation theory is used in the small parameter m / M (the ratio of the electron mass to the nuclear mass) [1]. In another version [2], the same problem is solved exactly.

2. Perturbation theory

Let us first dwell on work [1]. The nonrelativistic Schrödinger equation for a helium atom with a fixed nucleus has the form (in atomic units):

$$-\frac{1}{2}\Delta_1\Psi - \frac{1}{2}\Delta_2\Psi + \left(\frac{1}{r_{12}} - \frac{2}{r_2} - \frac{2}{r_1}\right)\Psi = E_0\Psi,$$

where the wave function $\Psi = \Psi(r_1, r_2)$ depends on the coordinates of two electrons r_1 and r_2, $r_i = |r_i|$, Δ_i is the Laplacian acting on r_i, $r_{12} = |r_1 - r_2|$, the atomic nucleus is at the origin coordinates. E_0 - eigenvalue of energy for the ground state. This value was calculated by a number of authors. From work [1] we have:

$$E_0 = -2.903724377034120.$$

The ground state energy, expanded in powers of *m/M*, is given in [1]:

$$E_g = E_0 + (m/M)E_1 + (m/M)^2E_2\,,$$

where $E_1 = 3.06279384$, $E_2 = -3.692271$. We will take the *m/M* value from [2]:

$$M/m = 7294.299620097.$$

Journal of Physics: Conference Series **2067** (2021) 012002 doi:10.1088/1742-6596/2067/1/012002

This value with the required accuracy (the first six digits) coincides with the CODATA recommendation: *M/m* = 7294.29954142 (24) [3]. Substituting the value of m/M in (1), we get:

$$E_g = -2.903304557748695.$$

3. Exact solution

In [2], a method was developed for calculating the nonrelativistic energy levels of a three-particle Coulomb system with arbitrary masses and charges. The Schrödinger equation for a system of three particles with Coulomb interaction is written as follows (in atomic units):

$$-\sum_{i=1}^{3} \frac{1}{2M_i} \Delta_i \Psi + \left(\frac{Z_1 Z_2}{r_{12}} + \frac{Z_2 Z_3}{r_{23}} + \frac{Z_3 Z_1}{r_{31}} \right) \Psi = E\Psi ,$$

M_i and Z_i - particle mass and charge, E - energy. For states with a total angular momentum equal to zero, the wave function does not depend on the angular variables, that is,

$$\Psi = \Psi(r_{12}, r_{23}, r_{31}) .$$

The coordinates $r_{12}, r_{23}, r_{31}$are not independent. This difficulty can be overcome by using the independent variables *u, v, w*, defined as follows:

$$u = r_{12} - r_{23} + r_{31}, \quad v = r_{23} - r_{31} + r_{12}, \quad w = r_{31} - r_{12} + r_{23}.$$

Variables *u, v, w* ranges from 0 to ∞. The wave function is sought in the form:

$$\Psi = exp(-(u + v + w)/2)\psi(u, v, w).$$

The function ψ is expanded in a series in Laguerre polynomials:

$$\psi(u, v, w) = \sum_{n,m,k=0}^{\infty} L_n(u) L_m(v) L_k(w) \psi_{nmk} .$$

Considering the properties of the Laguerre polynomials and the normalization condition

$$\int_0^{\infty} dx\, e^{-x} L_n(x) L_{n'}(x) = \delta_{nn'},$$

in [2], an exact recurrence relation was obtained for the coefficients ψ_{nmk}. This relationship is an infinite system of linear equations for finding the eigenvectors ψ_{nmk} and eigenvalues of energy *E*:

$$\sum_{n'm'k'=0}^{\infty} \langle n\, m\, k | C | n'm'k' \rangle\, \psi_{n'm'k'} = E \sum_{n'm'k'=0}^{\infty} \langle n\, m\, k | R | n'm'k' \rangle\, \psi_{n'm'k'} ,$$

where $\langle n\, m\, k | C | n'm'k' \rangle$ and $\langle n\, m\, k | R | n'm'k' \rangle$ known matrix elements. Each coefficient ψ_{nmk} is determined by three numbers, which is not very convenient for numerical solution of equations. Therefore, to each triple n, m, k we associate the index i, which ranges from 0 to ∞. This correspondence is established as follows. The number p = n + m + k is entered, where $p = 0, 1, \ldots$. For a fixed p, the numbers i are ordered as follows. First, n changes from 0 to p, then (for a fixed n) m changes from 0 to p-n, while k = p-n-m. Thus, we have:

Journal of Physics: Conference Series **2067** (2021) 012002 doi:10.1088/1742-6596/2067/1/012002

$$\sum_{j=0}^{\infty} C_{ji}\,\psi_i = E \sum_{j=0}^{\infty} R_{ji}\,\psi_i\,.$$

In work [2] this system was solved for the energy E by the method of successive approximations. A convergent sequence of energy values was obtained with an increase in the number of solved equations. In the case of a helium atom, the masses and charges of the particles are as follows:

$$M_1 = 1, \qquad M_2 = 1, \qquad M_3 = M,$$

$$Z_1 = -1, \qquad Z_2 = -1, \qquad Z_3 = 2,$$

where M is the mass of the nucleus of a helium atom in atomic units. For the ground state energy, the following value was obtained:

$$E = -2.903304557734390.$$

4. Comparison of calculation results

The difference between the exact E and the approximate E_g values of the energy of the ground state of the helium atom is the error in calculating E_g:

$$E - E_g = 0.000000000014305.$$

The results of calculations of E and E_g coincide with an accuracy of 10^{-10}, which confirms the high accuracy of both methods. Taking into account that the atomic frequency unit is equal to 6579683920.50 *MHz* [3], we have

$$E - E_g = 0.094\ MHz.$$

5. Conclusion

The comparison shows that the difference between the two results of calculating the ground state energy of the helium atom is about 100 kHz. However, it should be noted that work [2] was carried out in 2000, when the computing power was significantly lower than at present. The motivation for this article is related to our studies [4, 5], where we focus on 10 kHz accuracy in the study off resonant stimulated Raman scattering on the 2^3S-1^1S transition of the helium atom (62.6 nm).

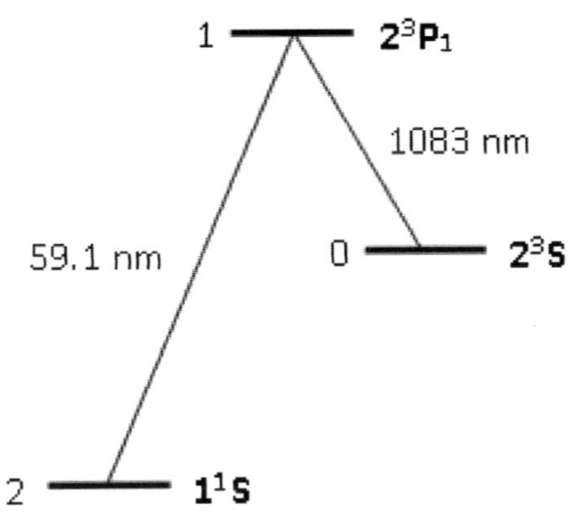

Journal of Physics: Conference Series **2067** (2021) 012002 doi:10.1088/1742-6596/2067/1/012002

The pump field frequency is close to the frequency of the $2^3P - 2^3S$ transition (1083 nm), and the stimulated scattering frequency is close to the frequency of the $2^3P - 1^1S$ transition (59.1 nm). The 2^3S-1^1S transition has a small radiative width, which is determined by the lifetime of the 2^3S state (7900 s). In [4], the possibility of measuring the frequency of the forbidden 2^3S-1^1S transition with an accuracy of 10 kHz was shown. Measuring the 2^3S-1^1S transition frequency with this precision will provide additional information for testing quantum electrodynamics. For example, you can measure the Lamb shift of the 2^3S level and compare it with the theoretical value (~ 5 GHz). The work [5] shows the possibility of creating a frequency standard based on the 2^3S-1^1S forbidden transition. The standard is needed to measure frequencies in the region of extreme ultraviolet radiation, and possibly X-ray radiation. This is relevant for research in chemistry and biochemistry, which are currently widely carried out using synchrotron radiation.

The study was carried out with the financial support of the Russian Foundation for Basic Research within the framework of scientific project No. 20-02-00068

6. References

[1] Yerokhin V A and Pachucki K 2010 *Phys. Rev.* A **81** 022507
[2] Baklanov E V and Denisov A V 2000 *Laser Physics* **10**, 1.
[3] CODATA physics.nist.gov/cuu/Constants/index.html, 2021
[4] Baklanov E V and Taichenachev A V 2018 *Opt. Spectrosc.* **124** 285
[5] Baklanov E V and Taichenachev A V 2020 *Opt. Spectrosc.* **128** 1715

Method for stabilization of the microwave modulation index in order to suppress the light shift of the coherent population trapping resonances

V I Yudin[1−3], M Yu Basalaev[1−3], A V Taichenachev[1,2], D A Radnatarov[1], V A Andryushkov[1] and S M Kobtsev[1]

[1] Novosibirsk State University, ul. Pirogova 1, Novosibirsk, 630090, Russia
[2] Institute of Laser Physics SB RAS, pr. Ak. Lavrent'eva 15B, Novosibirsk, 630090, Russia
[3] Novosibirsk State Technical University, pr. Karla Marksa 20, Novosibirsk, 630073, Russia

E-mail: viyudin@mail.ru

Abstract. For the resonance of coherent population trapping (CPT), we show that in the case of a spatially inhomogeneous light shift (for example, due to the Gaussian transversal profile of the light beam intensity), the zero position of the error signal, formed by the use of phase-jump technique, depends on the integration time of the spectroscopic signal. Basing on this effect, we propose two-loop method to stabilize the microwave power at the point where the light shift vanishes.

1. Introduction

Microwave atomic clocks based on coherent population trapping (CPT) resonance [1, 2, 3, 4, 5, 6] are attractive for such applications as secure communications, global satellite navigation and positioning systems. The advantages of these devices is a fully optical excitation scheme of narrow rf resonance without a microwave cavity. This enables one to significantly reduce the dimension of the physical package (up to the chip-scale) and power consumption.

Spectroscopic measurements during the frequency stabilization are accompanied by a perturbation of clock transition frequency due to the ac-Stark effect. This light shift depends on the optical field intensity and is one of the key factors limiting the metrological characteristics (stability and accuracy) of atomic clocks. Various methods have been suggested to suppress light shift in CPT clocks based on Ramsey [7, 8, 9, 10, 11, 12, 13] or continuous-wave [14, 15, 16] spectroscopy. However, these methods require the use of an external laser power modulator, which limits the possibility of their implementation in chip-scale atomic clocks.

In present paper, we develop an alternative two-loop method for suppressing light shift in continuous-wave spectroscopy of CPT resonance in the case of spatially nonuniform profile of the light intensity. Our approach based on phase-jump spectroscopy [17] and power stabilization of the microwave signal (which is used to modulate the VCSEL current). The proposed method does not need a laser power modulator and is free from the problem of lineshape-asymmetry-induced shift (see more detail in Ref. [17]).

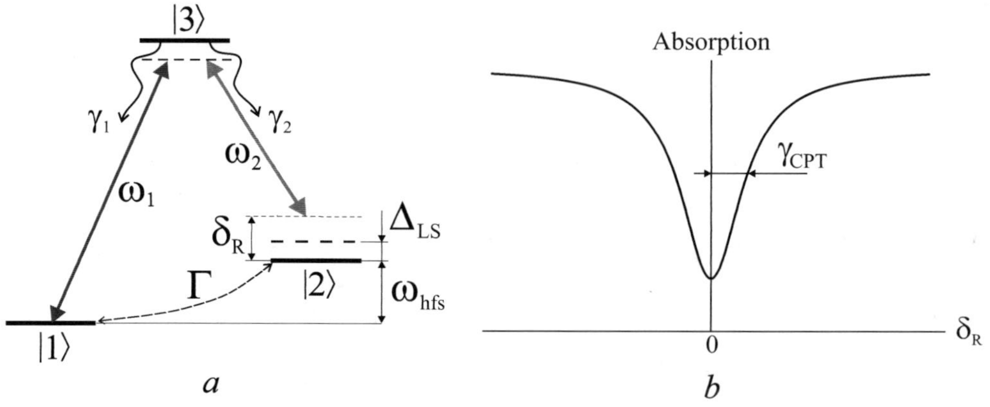

Figure 1. (a) The scheme of the three-level Λ system, where ω_1 and ω_2 are the frequencies of resonant optical fields; $\Delta_{\rm sh}$ is light (Stark) shift of the clock transition frequency; γ_1 and γ_2 are the rates of spontaneous population transfer from upper state $|3\rangle$ to the lower states $|1\rangle$ and $|2\rangle$, respectively; Γ is decay rate of the coherence between the states $|1\rangle$ and $|2\rangle$. (b) CPT resonance in the absorption spectrum, where $\gamma_{\rm CPT}$ is half-width at half-maximum.

2. Theoretical model

As a theoretical model of the atomic medium, we consider a closed three-level Λ system (see figure 1a) interacting with a bichromatic light field:

$$E(t) = E_1 e^{-{\rm i}[\omega_1 t+\varphi_1(t)]} + E_2 e^{-{\rm i}[\omega_2 t+\varphi_2(t)]} + \text{c.c.}, \tag{1}$$

under conditions of coherent population trapping, that is, when the difference of frequencies $\omega_1 - \omega_2$ is scanned at the frequency $\omega_{\rm hfs}$ of the transition between the lower states of the Λ system. We will describe the time dynamics of the Λ system using the formalism of an atomic density matrix, which has the following form in the basis of states $\{|j\rangle\}$:

$$\hat{\rho}(t) = \sum_{m,n} |m\rangle \rho_{mn}(t) \langle n|. \tag{2}$$

In the rotating wave approximation the density matrix elements satisfy the following equations:

$$\begin{aligned}
\partial_t \rho_{31} &= (-\gamma_{\rm opt} + {\rm i}\delta_1)\,\rho_{31} + {\rm i}\Omega_2 e^{-{\rm i}\varphi_2(t)}\rho_{21} + {\rm i}\Omega_1 e^{-{\rm i}\varphi_1(t)}\,(\rho_{11} - \rho_{33}),\\
\partial_t \rho_{32} &= (-\gamma_{\rm opt} + {\rm i}\delta_2)\,\rho_{32} + {\rm i}\Omega_1 e^{-{\rm i}\varphi_1(t)}\rho_{12} + i\Omega_2 e^{-{\rm i}\varphi_2(t)}\,(\rho_{22} - \rho_{33}),\\
\partial_t \rho_{21} &= [-\Gamma + {\rm i}\,(\delta_{\rm R} - \Delta_{\rm LS})]\,\rho_{21} - {\rm i}\Omega_1 e^{-{\rm i}\varphi_1(t)}\rho_{23} + {\rm i}\Omega_2 e^{{\rm i}\varphi_2(t)}\rho_{31},\\
\partial_t \rho_{11} &= -\Gamma\,(\rho_{11} - 1/2) + \gamma_1\rho_{33} - {\rm i}\Omega_1 e^{-{\rm i}\varphi_1(t)}\rho_{13} + {\rm i}\Omega_1 e^{{\rm i}\varphi_1(t)}\rho_{31},\\
\partial_t \rho_{22} &= -\Gamma\,(\rho_{22} - 1/2) + \gamma_2\rho_{33} - {\rm i}\Omega_2 e^{-{\rm i}\varphi_2(t)}\rho_{23} + {\rm i}\Omega_2 e^{{\rm i}\varphi_2(t)}\rho_{32},\\
\partial_t \rho_{33} &= -(\gamma_{\rm sp} + \Gamma)\rho_{33} + {\rm i}\Omega_1 e^{-{\rm i}\varphi_1(t)}\rho_{13} - {\rm i}\Omega_1 e^{{\rm i}\varphi_1(t)}\rho_{31} + {\rm i}\Omega_2 e^{-{\rm i}\varphi_2(t)}\rho_{23} - {\rm i}\Omega_2 e^{{\rm i}\varphi_2(t)}\rho_{32},\\
\rho_{12} &= \rho_{21}^*, \quad \rho_{13} = \rho_{31}^*, \quad \rho_{23} = \rho_{32}^*,
\end{aligned} \tag{3}$$

under the condition of normalization (conservation of the total population):

$$\text{Tr}\{\rho\} = \rho_{11} + \rho_{22} + \rho_{33} = 1. \tag{4}$$

In equations (3) we use the following notations: $\Omega_1 = |d_{31}E_1|/\hbar$ and $\Omega_2 = |d_{32}E_2|/\hbar$ are Rabi frequencies for the transitions $|1\rangle \leftrightarrow |3\rangle$ and $|2\rangle \leftrightarrow |3\rangle$, respectively ($d_{31}$ and d_{32} are matrix

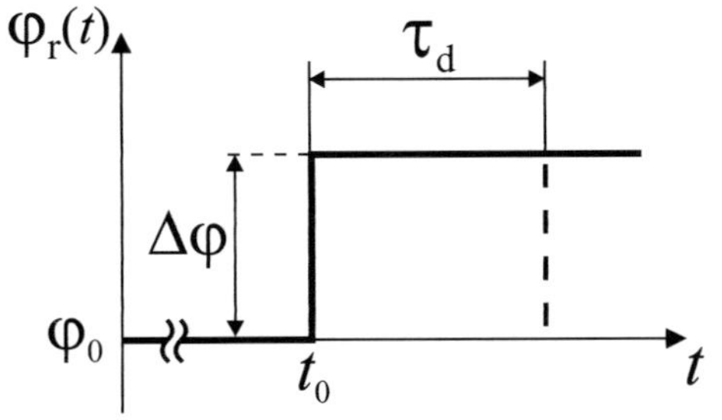

Figure 2. Scheme of jump modulation of the relative phase $\varphi_r = \phi_1 - \phi_2$ of the bichromatic field, φ_0 is initial phase difference of fields under stationary pumping of atoms, τ_d is detection time of the signal.

elements of the operator of electric dipole interaction); $\delta_1 = \omega_1 - \omega_{31}$ and $\delta_2 = \omega_2 - \omega_{32}$ are one-photon detunings of laser spectral components; $\delta_R = \delta_1 - \delta_2 = \omega_1 - \omega_2 - \omega_{hfs}$ is two-photon (Raman) detuning; Δ_{LS} is the light (Stark) shift of the clock transition frequency; γ_{opt} is the damping rate of optical coherence (due to spontaneous decay processes, collisions with buffer gas atoms, etc.); γ_1 and γ_2 are the rates of spontaneous population transfer from a state $|3\rangle$ to states $|1\rangle$ and $|2\rangle$, respectively; $\gamma_{sp} = \gamma_1 + \gamma_2$ is the spontaneous decay rate of the excited state $|3\rangle$; and the constant Γ describes the relaxation of atoms (for instance, due to transit effects) to an equilibrium isotropic distribution over the lower energy levels of the Λ system.

As a spectroscopic signal, we will study the absorbed power which, in the case of an optically thin medium, is proportional to

$$A(t) = \frac{\partial \rho_{33}}{\partial t} + (\gamma_{sp} + \Gamma)\rho_{33}. \tag{5}$$

Let us extract in the Rabi frequencies and light shift the spatial dependence of the transverse profile of the light beam f on the radius r:

$$\Omega_1^2(r) = \Omega_{10}^2 f(r), \quad \Omega_2^2(r) = \Omega_{20}^2 f(r), \quad \Delta_{LS}(r) = \Delta_0 f(r), \tag{6}$$

where Ω_{10}, Ω_{20} and Δ_0 are the Rabi frequencies and the light shift on the axis (i.e. at $r = 0$) of the light beam, respectively. In the case of a Gaussian light beam profile, the function $f(r)$ has the form

$$f(r) = e^{-r^2/r_0^2}, \tag{7}$$

where the radius r_0 determines the transverse size of the beam. To simulate the spectroscopic signal from the photodetector, it is necessary to integrate the local absorption (5) over the transverse profile of the laser beam:

$$\bar{A}(t) = \int_0^\infty \int_0^{2\pi} A(r,t) r d\varphi dr. \tag{8}$$

3. Phase-jumps spectroscopy

In this section, we briefly describe the main idea of the phase-jumps spectroscopy [17]. In this approach, the excitation of the dynamic (that is, time-dependent) response of the quantum system is carried out by modulating the relative phase of the bichromatic field

$$\varphi_r(t) = \varphi_1(t) - \varphi_2(t) \tag{9}$$

according to the jump law:

$$\varphi_r(t) = \begin{cases} \varphi_0, & \text{if } t < t_0; \\ \varphi_0 + \Delta\varphi, & \text{if } t \geq t_0, \end{cases} \tag{10}$$

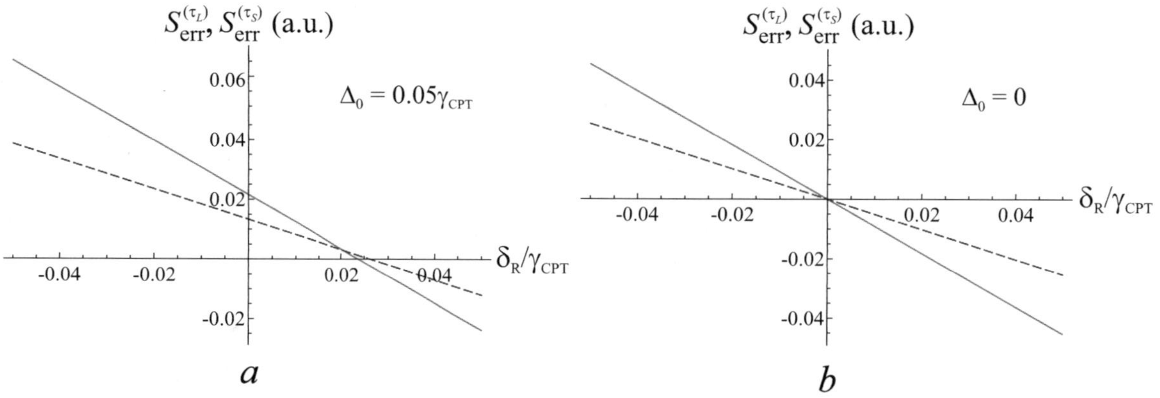

Figure 3. Error signals (11) for different detection times $\tau_L = 5\,\gamma_{\rm CPT}^{-1}$ (solid red line) and $\tau_S = 1.5\,\gamma_{\rm CPT}^{-1}$ (dashed blue line):
(a) CPT resonance is shifted, $\Delta_0 = 0.05\,\gamma_{\rm CPT}$;
(b) light shift suppressed, $\Delta_0 = 0$.
Calculation parameters: $\Omega_{10}^2 + \Omega_{20}^2 = 0.01$, $\Omega_{10}/\Omega_{20} = 1.2$, $\gamma_{\rm opt} = 50\gamma_{\rm sp}$, $\gamma_1 = \gamma_2 = \gamma_{\rm sp}/2$, $\Gamma = 10^{-4}\gamma_{\rm sp}$, $\delta_1 = \delta_2 = 0.1\,\gamma_{\rm opt}$.

where φ_0 is the initial phase difference of the two-frequency field, and $\Delta\varphi$ is the value of phase jump. A scheme of modulation of the relative phase $\varphi_{\rm r}$ is shown in figure 2. First the atoms are pumped to a steady state. Then the relative phase changes, in a jump, by some value, which causes a transition process in the absorption signal. At an exact two-photon resonance ($\delta_{\rm R} = 0$) the dynamics of the transition process does not depend on the sign of the jump. However, at detuning from resonance ($\delta_{\rm R} \neq 0$) the time evolution of the spectroscopic signal becomes different for phase jumps of opposite signs. The error signal for frequency stabilization can be generated using phase jumps of different signs ($\pm\Delta\varphi$) as follows:

$$S_{\rm err}(\delta_{\rm R}) = \int_{t_0}^{t_0+\tau_{\rm d}} \bar{A}(t, +\Delta\varphi)dt - \int_{t_0}^{t_0+\tau_{\rm d}} \bar{A}(t, -\Delta\varphi)dt, \quad (11)$$

where t_0 is the start time of the spectroscopic signal recording; $\tau_{\rm d}$ is the duration of the signal accumulation (detection time). Thus, the error signal is defined as the difference between the area below the absorption curve for a positive jump of the relative phase and the area below the absorption curve for a negative phase jump. The frequency stabilization of the local oscillator is near the zero error signal: $S_{\rm err}(\delta_{\rm R}) = 0$. To maximize the error signal slope we need to use $\pm\pi/2$ phase jumps.

4. Method for stabilizing the microwave modulation index

In compact (including chip-scale) CPT atomic clocks, vertical-cavity surface-emitting laser (VCSEL) are used to generate a multi-frequency field. Harmonic modulation of the laser current using a signal from a microwave generator leads to phase modulation of the laser radiation, which corresponds to a set of equidistant coherent frequencies (harmonics) in the radiation spectrum. The modulation index is determined by the power of the microwave signal. The main contribution to the formation of the CPT state is introduced by two resonant field components. Nonresonant harmonics lead to a shift of the clock transition frequency due to the ac Stark effect. However, it is possible to select the modulation index of the laser radiation so that light shift is absent (see, for example, [18, 14, 19]).

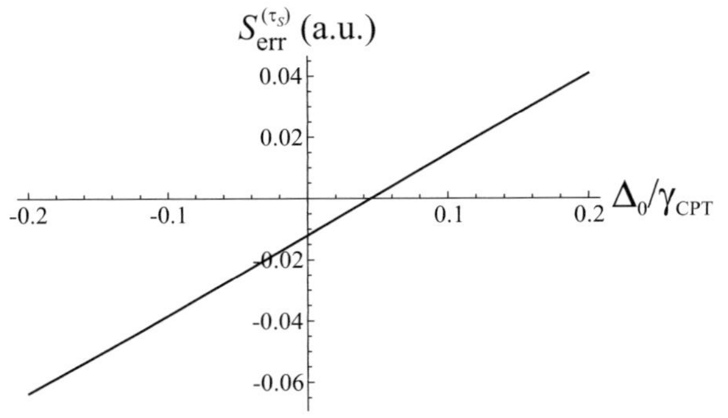

Figure 4. The error signal (11) with the short integration time τ_S as a function of the light shift at the center of the Gaussian profile Δ_0. Calculation parameters: $\Omega_{10}^2+\Omega_{20}^2 = 0.01$, $\Omega_{10}/\Omega_{20} = 1.2$, $\gamma_{\text{opt}} = 50\gamma_{\text{sp}}$, $\gamma_1 = \gamma_2 = \gamma_{\text{sp}}/2$, $\Gamma = 10^{-4}\gamma_{\text{sp}}$, $\delta_1 = \delta_2 = 0.1\,\gamma_{\text{opt}}$, $\tau_l = 5\,\gamma_{\text{CPT}}^{-1}$, $\tau_s = 1.5\,\gamma_{\text{CPT}}^{-1}$, δ_{R} is a solution to the equation $S_{\text{err}}^{(\tau_L)}(\delta_{\text{R}}) = 0$ at $\Delta_0 = 0.05\,\gamma_{\text{CPT}}$.

Using in the expression (11) two different integration times, long ($\tau_{\text{d}} = \tau_L$) and short ($\tau_{\text{d}} = \tau_S < \tau_L$), we can generate two error signals: $S_{\text{err}}^{(\tau_L)}(\delta_{\text{R}})$ and $S_{\text{err}}^{(\tau_S)}(\delta_{\text{R}})$, respectively. Figure 3 demonstrate the error signals calculated for $\tau_L = 5\,\gamma_{\text{CPT}}^{-1}$ and $\tau_S = 1.5\,\gamma_{\text{CPT}}^{-1}$, where γ_{CPT} is the half-width at half-maximum of the CPT resonance (see figure 1*b*). As can be seen from figure 3*a*, in the case of the light shift of the clock transition frequency, the zeros of the error signals are located at different two-photon detunings. Whereas in the absence of the light shift, the zeros of the error signals coincide, as shown in figure 3*b*. This feature can be used to organize a second feedback loop, which will stabilize the power of the microwave signal at a value where the light shift vanishes.

We propose the following algorithm for stabilizing the frequency of the local oscillator and the power of the microwave generator. For the error signal with the long integration time τ_L, the microwave power is fixed (that is, $P_{\text{mw}} = P_{\text{mw}}^{(fix)}$), and the variable detuning δ_{R} is stabilized at the zero of the first error signal:

$$S_{\text{err}}^{(\tau_L)}(\delta_{\text{R}}, P_{\text{mw}}^{(fix)}) = 0. \tag{12}$$

Then, measurements are carried out for the error signal with the short integration time τ_S, where the two-photon detuning found in the previous step is fixed (i.e. $\delta_R = \delta_{\text{R}}^{(fix)}$) and the microwave power P_{mw} is stabilized at the zero of the second error signal:

$$S_{\text{err}}^{(\tau_S)}(\delta_{\text{R}}^{(fix)}, P_{\text{mw}}) = 0. \tag{13}$$

Figure 4 shows the error signal with a short integration time $S_{\text{err}}^{(\tau_S)}$ as a function of the light shift Δ_0 at the center of the Gaussian profile determined by the microwave power P_{mw}. Repeating these iterations, we stabilize both parameters $\delta_R = \bar{\delta}_{\text{R}}$ and $P_{\text{mw}} = \bar{P}_{\text{mw}}$ which correspond to the solution of the following equations system:

$$S_{\text{err}}^{(\tau_L)}(\bar{\delta}_{\text{R}}, \bar{P}_{\text{mw}}) = 0, \quad S_{\text{err}}^{(\tau_S)}(\bar{\delta}_{\text{R}}, \bar{P}_{\text{mw}}) = 0. \tag{14}$$

In this case, our calculations show (see figure 3*b*) that $\bar{\delta}_{\text{R}} = 0$, i.e. the light shift of the stabilized clock frequency is absent.

5. Conclusion

It is shown that in the case of phase-jump spectroscopy of CPT resonance, the spatial inhomogeneity of the light shift (for example, due to the Gaussian transversal intensity profile of the light beam) leads to the dependence of the zero-point of the error signal on the integration time. We propose to use this feature for active suppression of light shift by adjusting the microwave power. The criterion for the absence of a light shift is the condition when the zeros of the two error signals, generated at different integration times, coincide.

Acknowledgments
This work was supported by the Russian Science Foundation (grant No. 21-12-00057).

References
[1] Vanier J 2005 *Appl. Phys. B* **81** 421
[2] Shah V and Kitching J 2010 *Adv. At. Mol. Opt. Phys.* **59** 21
[3] Knappe S 2008 *in Comprehensive Microsystems* (Oxford: Elsevier) 571-612
[4] Wang Z 2014 *Chin. Phys. B* **23** 030601
[5] Kitching J 2018 *Appl. Phys. Rev.* **5** 031302
[6] Carlé C, Petersen M, Passilly N, Abdel Hafiz M, de Clercq E and Boudot R 2021 *IEEE Trans Ultrason Ferroelectr Freq Control* **68** 3249
[7] Yudin V I, Taichenachev A V, Basalaev M Yu, Zanon-Willette T, Pollock J W, Shuker M, Donley E A and Kitching J 2018 *Phys. Rev. Applied* **9** 054034
[8] Yudin V I, Taichenachev A V, Basalaev M Yu, Mehlstäubler T E, Boudot R, Zanon-Willette T, Pollock J W, Shuker M, Donley E A and Kitching J 2018 *New J. Phys.* **20** 123016
[9] Abdel Hafiz M, Coget G, Petersen M, Rocher C, Guérandel S, Zanon-Willette T, de Clercq E and Boudot R 2018 *Phys. Rev. Applied* **9** 064002
[10] Abdel Hafiz M, Coget G, Petersen M,Calosso C E, Guérandel S, de Clercq E and Boudot R 2018 *Appl. Phys. Lett.* **112** 244102
[11] Shuker M, Pollock J W, Boudot R,Yudin V I, Taichenachev A V, Kitching J and Donley E A 2019 *Phys. Rev. Lett.* **122** 113601
[12] Shuker M, Pollock J W, Boudot R, Yudin V I, Taichenachev A V, Kitching J and Donley E A 2019 *Appl. Phys. Lett.* **114** 141106
[13] Basalaev M Yu, Yudin V I, Kovalenko D V, Zanon-Willette T and Taichenachev A V 2020 *Phys. Rev. A* **102** 013511
[14] Shah V, Gerginov V, Schwindt P D D, Knappe S, Hollberg L and Kitching J 2006 *Appl. Phys. Lett.* **89** 151124
[15] Yudin V I, Basalaev M Yu, Taichenachev A V, Pollock J W, Newman Z L, Shuker M, Hansen A, Hummon M T, Boudot R, Donley E A and Kitching J 2020 *Phys. Rev. Applied* **14** 024001
[16] Abdel Hafiz M, Vicarini R, Passilly N, Calosso C E, Maurice V, Pollock J W, Taichenachev A V, Yudin V I, Kitching J and Boudot R 2020 *Phys. Rev. Applied* **14** 034015
[17] Basalaev M Yu, Yudin V I, Taichenachev A V, Vaskovskaya M I, Chuchelov D S, Zibrov S A, Vassiliev V V and Velichansky V L 2020 *Phys. Rev. Applied* **13** 034060
[18] Zhu M and Cutler L S 2000 *Proceedings of 32th Annual Precise Time and Time Interval Systems and Applications Meeting* (Reston: Institute of Navigation) p 311
[19] Yudin V I, Taichenachev A V and Basalaev M Yu 2016 *Phys. Rev. A* **93** 013820

Journal of Physics: Conference Series **2067** (2021) 012004 doi:10.1088/1742-6596/2067/1/012004

Effects of size and dimension in optic relaxation in Rb vapor

S Atutov and V Sorokin

Institute of Automation and Electrometry SB RAS, Koptyug Ave. 1, 630090 Novosibirsk, Russia
E-mail: vlad_sorokin@ngs.ru

Abstract. We report on an experiment in which the fluorescence decay time of 5P levels of Rb atoms in a coated vapor cell exceeds several of milliseconds that many orders of magnitude longer than normal decay time of excited of rubidium atoms. This effect has been observed by using the pump laser with fixing frequency and the periodically scan the probing laser frequency of which overlap all Rb hyperfine linens of 5P-5S transition. Long time fluorescence decay take place in wide part of the spherical high quality coated cell. In cylindrical part of the cell or in uncoated cell the long time fluorescence decay was disappeared.

The study of the optical properties of rubidium vapors has recently acquired considerable relevance due to the prospects for creating precision atomic time standards [1-3], as well as other quantum devices using long-lived polarized states of rubidium atoms [4, 5]. In order to increase the lifetime of polarized states, which can be destroyed in collisions of atoms with the walls of the cell, various antirelaxation coatings based on saturated hydrocarbons are used: paraffin, ceresin [6]. Recently, coatings have been discovered that further increase the lifetime of polarized states of alkali metals based on high molecular weight alkenes - hydrocarbons with one double bond.

When creating an atomic clock, both the method of double radio-optical resonance and bichromatic laser radiation can be used. Under certain conditions, bichromatic laser radiation is capable of creating a coherent population trapping (CPT) resonance [7-9] between the ground state sublevels, split due to the interaction of the nuclear spin with the orbital moments and electron spin. The main condition for CPT resonance is the coincidence of the frequency difference of the bichromatic laser radiation with the hyperfine splitting of the ground state, which is the frequency reference for atomic frequencies. A CPT resonance is observed by the phenomenon of induced optical transparency, since CPT produces a coherent superposition of hyperfine sublevels of the ground state that does not interact with bizromatic laser radiation.

Of interest are other aspects of such interaction without the purposeful creation of a CPT. Earlier in [10], a long decay of the fluorescence of rubidium vapor in a spherical cell of a special design with a ceresin coating, which does not have a source of rubidium vapor with a free metal surface, was found [11]. In [10], one sufficiently intense laser pumping radiation had a fixed frequency, resonant to one of the transitions between the hyperfine components of the ground state ($5^2S_{1/2}$) and the excited state ($5^2P_{1/2}5^2P_{1/2}$) in a natural mixture of ^{85}Rb and ^{87}Rb vapors. The second intense probe laser radiation was frequency modulated according to a triangular law. The frequency modulation range was about 9 GHz, overlapping the spectral interval in which all allowed electric dipole transitions between the hyperfine sublevels of the $5^2P_{1/2}$ and $5^2S_{1/2}$ states are located. The frequency modulation period could be adjusted from 3 to 10 milliseconds. The long decay of the fluorescence of rubidium vapor was superimposed on the fluorescence spectrum and started at the time when the pump and probe radiation were in resonance with the three-level Λ level scheme, with the upper $5^2P_{1/2}$ state and the lower levels formed due to hyperfine splitting of the ground state.

In this paper, we present the results of a series of experiments similar to [10], but carried out with different configurations of laser beams with respect to the cell. The pump laser radiation frequency was chosen to be resonant to the $5^2P_{1/2}$–$5^2S_{1/2}$ (F = 3) transition of the ^{85}Rb isotope. Figure 1a shows a scheme for measuring the fluorescence signal, in which the cell with rubidium vapor consists of a spherical part 12 cm in diameter and a relatively thin tube 8 mm in diameter and 10 cm long. The cell is covered with ceresin from the inside. A piece of metallic rubidium is placed at the end of the tube farthest from the ball. The pump radiation propagates from right to left from the side of the ball. Probing radiation propagates in parallel to the pump radiation and towards it. The graphs of the fluorescence signal depending on the observation time associated with the frequency of the probe radiation with frequency modulation according to the triangular law for two registration zones are shown in figure 1b. In fact, figure 1b shows the fluorescence plots for half the period of frequency modulation of the probe radiation. The capillary zone (zone 1 in figure 1a) corresponds to graph 1 in figure 1b. For the center of the spherical part (zone 2 in figure 1a), the fluorescence signal is shown in graph 2 in figure 1b.

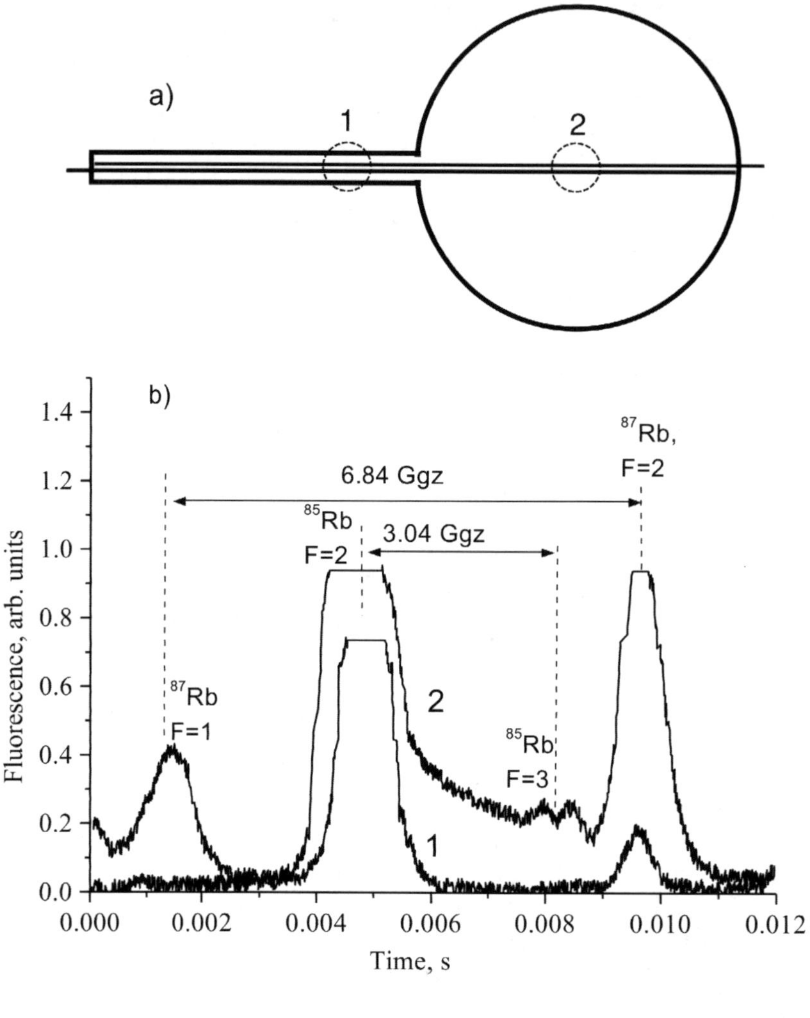

Figure 1. A scheme for measuring the fluorescence signal (a), fluorescence signal from registration zones 1 and 2 of the cell.

The fluorescence peaks in the region of the capillary 1 reproduce the shape of the absorption lines taking into account the optical pumping for each of the rubidium isotopes. In the center of the spherical part of the cuvette, the time dependence of the fluorescence signal is qualitatively different (graph 2 in figure 1b). In addition to the fluorescence peaks associated with the shape of the absorption lines, a slowly decaying fluorescence signal is clearly observed when the probe radiation has ceased to be resonant to the $5^2P_{1/2}-5^2S_{1/2}$ (F = 3) transition of the ^{85}Rb isotope, starting from the time interval 0.006 sec. In addition, a permanent backing appears in the fluorescence signal, which is apparently associated with the overlapping of slowly decaying fluorescence signals from different periods of frequency modulation of the probe radiation. Those. the slowly decaying fluorescence signal starts when the probe radiation is resonant with the $5^2P_{1/2}-5^2S_{1/2}$ (F = 2) transition of the ^{85}Rb isotope and lasts much longer than the half-period of frequency modulation, the duration of which in this experiment is about 0.012 sec. Graphs ha fig. 1b have truncated tops due to saturation of the recording system, the gain in which was increased in order to emphasize the amplitude of the slowly decaying fluorescence signal.

The fluorescence spectrum of zone 2 (graph 3 in figure 2b) with a closed pump beam represents four weak, at the level of noise, peaks characteristic of the absorption spectrum of rubidium isotopes, which appeared, apparently, from the action of the residual scattered probe radiation. The same mechanism of the appearance of weak peaks outside the zone of action of the probe radiation is also manifested in graph 2 in figure 2b. Graphs 1 to 3 in figure 2b are spaced vertically to better distinguish the spectral-temporal features of the fluorescence signal of rubidium isotopes.

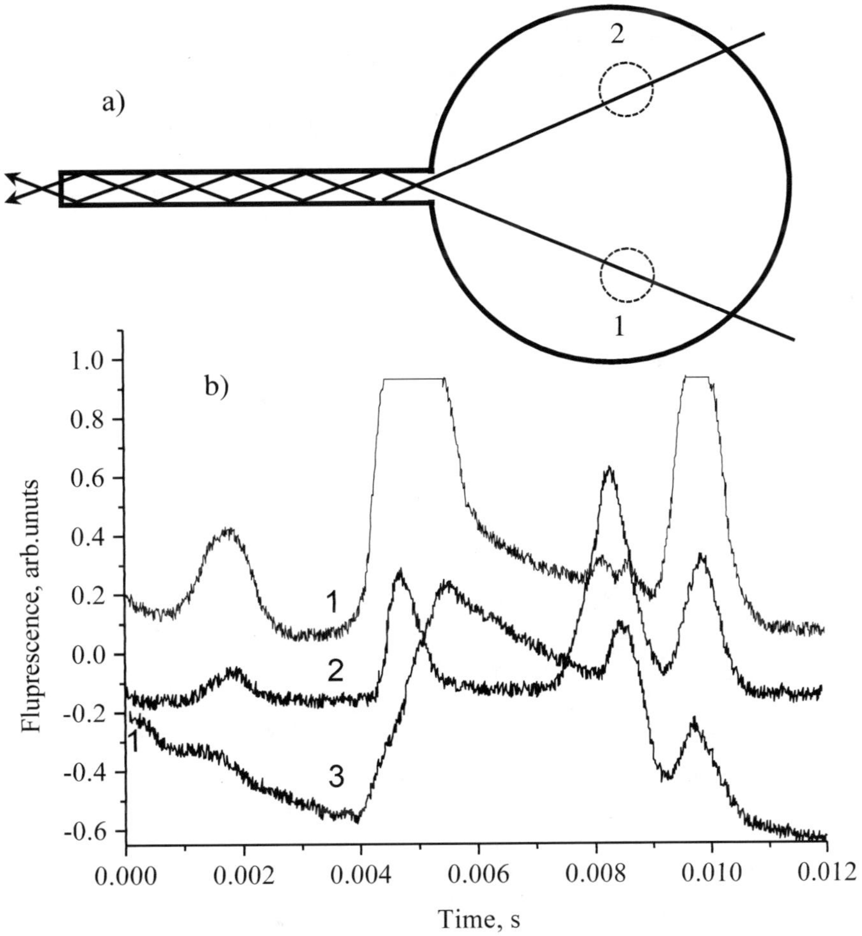

Figure 2. Fluorescence signal from registration zones 1 and 2 of the cell (b) with oblique propagation of the probe laser beam and the pump beam (a). Frequency scale is the same as in figure 1.

The experimental results demonstrate the dependence of the dynamics of relaxation processes in rubidium vapor on the size and dimension from registration scheme of the fluorescence signal. In the one-dimensional case (figure 1, registration zone 1), long decay time of the fluorescence is not observed. The growth of the decay time of fluorescence can be used as the basis for a simple method for determining the quality of the coating of the cells.

References

[1] Robinson H and Johnson C 1982 *Appl. Phys. Lett.* **40** 771
[2] Risley A, Jarvis S, and Vanier J 1980 *J. Appl. Phys.* **51** 4571
[3] Rahman C and Robinson H 1987 *IEEE J. Quantum Electron.* **23** 452
[4] Budker D and Romalis M 2007 *Nature Phys.* **3** 227.
[5] Smullin S, Savukov I, Vasilakis G, et al. 2009 *Romalis, Phys. Rev.* A **80** 033420
[6] Alexandrov E,. Balabas M, Budker D, et al. 2002 *Phys. Rev.* A **66** 042903
[7] Arimondo E, Orriols G 1976 *Lettere Al Nuovo Cimento* **17** 333
[8] Orriols G 1979 *Nuovo Cimento B* **53** 1
[9] Agapiev B, Gorny M, Matisov B, et al 1993 *UFN* **163** 1
[10] Atutov S., Sorokin V, Bagayev S, et al 2019 *Phys. J. D* **73** 240
[11] Atutov S, Plekhanov A, Sorokin V, et al 2018 *Eur. Phys. J. D* **72** 155

Journal of Physics: Conference Series **2067** (2021) 012005 doi:10.1088/1742-6596/2067/1/012005

Forced mode locking in a semiconductor laser

V F Zakharyash, V M Klement'ev and E A Titov

Institute of Laser Physics SB RAS, 15B Lavrentjeva Ave., Novosibirsk, Russia

Abstract. A brief description of the theory of mode locking of a semiconductor laser when the modulating generator frequency is close to the intermode frequency is given. It is shown that during mode locking there is phase locking, due to which the dispersion of summed intensity in the interpulse interval tends to zero.

e-mail: titov@laser.nsc.ru

As noted in a number of papers, a diode laser in long cavity is a convenient tool for studying the process of forced mode locking owing to relatively simple ways of obtaining locking modes [1, 2] and the possibility of choosing parameters of the external cavity with a convenient intermode frequency [3]. In such system a multi-mode regime can be observed for eigen frequency range. Diode current modulation with a frequency close to the intermode one results in mode locking. The intermode coupling required for it emerges because cavity modes along the diode are non-orthoganal.
The procedure of obtaining forced mode locking should be divided into two stages. The first stage is the free running mode (preparation, tuning, etc.), and the second stage is obtaining of an active or forced mode locking.

At the beginning of forced locking φ a redistribution of the noise is observed. This is mostly due to the pulling effect, when there is phase alignment of the modes ($\varphi_1 = \varphi_2 = \varphi_3$ =etc.), which is confirmed experimentally.
The theoretical study yields an interesting result. We have a multimode operation. The total field strength is written as

$$\mathrm{E}(z,t) = \varepsilon_0 \sum_k \sqrt{\xi_k} e^{i\varphi_k - i\omega_k t + ik_k z}, \tag{1}$$

Where $\varepsilon_0 = \sqrt{2\pi\hbar\omega / V}$ is the photon field strength. For simplicity, consider a ring cavity. The fields of individual modes are in a coherent state
$a_k |\alpha_k\rangle = \alpha_k |\alpha_k\rangle, \alpha_k = \sqrt{\xi_k} e^{i\varphi_k}$, where a_k is the annihilation operator for the k-th mode. The stationary distribution function for the k-th mode can be presented as

$$P(\xi_k) = \frac{1}{\sqrt{2\pi\Delta_k}} \exp\left[-(\xi_k - n_k)^2 / 2\Delta_k\right], \tag{2}$$

Where n_k is the average number of photons in a mode, and Δ_k is the variance of the photon distribution function in the mode, $\Delta_k = n_k d_k, d_k$ – connected with media's noise [4].
The distribution for the total field intensity is presented in the form [5]

$$W(|E|^2) = \int \delta(|E|^2 - \left|\sum_k \sqrt{\xi_k} e^{i\varphi_k - i\omega_k t}\right|^2) \prod_k^N P(\xi_k) d\xi_k d\varphi_k / (2\pi)^N . \quad (3)$$

In formula (3), the intensity is taken to be dimensionless ($\varepsilon_0 = 1$) and is sought for at point z=0.
In the absence of mode locking averaging over φ_k yields

$$W(|E|^2) = \int \delta(|E|^2 - \sum_k \xi_k) \prod_k^N P(\xi_k) d\xi_k . \quad (4)$$

Presenting the delta function in the form

$$\delta(|E|^2 - \sum_k \xi_k) = \frac{1}{2\pi} \int dp e^{ip|E|^2 - ip\sum_k \xi_k} ,$$

substituting into (4) and integrating by ξ_k and by p, we obtain

$$W(|E|^2) = \frac{1}{\sqrt{2\pi\Delta}} \exp\left[-(|E|^2 - \overline{n})^2 / 2\Delta\right], \quad (5)$$

Where $\overline{n} = \sum_k n_k , \Delta = \sum_k \Delta_k = \sum_k n_k d_k$.

In the case of mode locking $\varphi_k = \varphi_l$ for all k and l

$$|E|^2 = \sum_{k,l} \sqrt{\xi_k \xi_l} e^{-i(\omega_k - \omega_l)t} .$$

We set $\omega_k = \omega_0 + k\delta$, where k varies from -N/2 to N/2 (N is the number of modes).For simplicity, the amplitudes of all modes are assumed to be approximately the same.
Then for $|E|^2$ we have [6]

$$|E|^2 = \xi f(t), f(t) = \frac{\sin^2 \frac{N\delta t}{2}}{\sin^2 \frac{\delta t}{2}}, \; f(t) = \left| \sum_{k=-N/2}^{N/2} e^{-i\delta kt} \right|^2 . \quad (6)$$

The expression for $W(|E|^2)$ is as follows:

$$W(|E|^2) = \frac{1}{2\pi} \int dp e^{ip|E|^2} \int d\xi e^{-(\xi - n)^2/2\Delta} / \sqrt{2\pi\Delta} e^{-ip\xi f} , \quad (7)$$

Where n is the average number of photons in one mode, and $\Delta = nd$ – is the variance for one mode.
Integrating in (7), we obtain

$$W(|E|^2) = \frac{1}{\sqrt{2\pi\Delta f}} \exp\left[-(|E|^2 - nf)^2 / 2f^2\Delta\right]. \quad (8)$$

With (8), for the variance of the number of photons in the total field we have: $|E|^2 = n(t)$

$$\left\langle \left(\Delta n(t)\right)^2 \right\rangle = nf(t)(1 + df(t)) \quad (9)$$

Thus, outside the pulse the variance for the total field is zero. In the process of synchronization, not only modes of field summed into pulse, but noise is also summed. This gives a time dependent variance of total intensity of field (9).

References

[1] Matveev A N, Kolachevsky N N, Alnis J and Hänsch T 2008 *Quant. Electron.* **38** n 4 391
[2] Zakhar'yash V F, Kashirskii A V , Klement'ev V M 2015 Active mode-locking in semiconductor lasers Optoelectronics, *Instrumentation and Data Processing*. **51** n 6 553
[3] Klement'ev V M , Korel I I, Kurbatov A A, Titov E A 2014 Semiconductor diode in a long cavity in the active mode-locking regime *Optics and Spectroscopy*. **116** n 2 291
[4] Kazantsev A P, Surdutovich G I 1969 The Quantum Theory of the Laser Soviet Journal of *Experimental and Theoretical Physics*. **29** n 6 1075
[5] Glauber R. 1966 *Quantum Optics and Quantum Radiophysics (Moscow: Mir)*. p. 231. (In Russian)
[6] Baklanov EV 1980 *Physical foundations of the theory of lasers (Novosibirsk)*. p. 89. (In Russian)

Light shifts in the rubidium CPT atomic clock with laser current modulation at 3.4 and 6.8 GHz

S M Ignatovich[1], I S Mesenzova[1], M N Skvortsov[1], N L Kvashnin[1] and V I Vishnyakov[1].

[1] Institute of Laser Physics SB RAS, 15 B, Ac. Lavrentieva ave., Novosibirsk, 630090, Russia

stepani@ngs.ru

Abstract. In the paper presents results of an experimental comparison of the CPT parameters of the resonance at the D1 line in ^{87}Rb when the laser current is modulated at frequencies 3.4 GHz and 6.8 GHz. Resonances are investigated for slope to noise ratio (Q-factor) and shifts from the microwave power. The reproducibility of the parameters of VCSEL lasers required to obtain long-term stability of an atomic clock is considered. The obtained CPT instabilities of the atomic clock for 1 second were $1.2 \cdot 10^{-11}$ for the case of 3.4 GHz and $4 \cdot 10^{-12}$ for 6.8 GHz.

1. Introduction.

The coherent population trapping (CPT) resonance in alkali atoms is widely used to build miniature atomic clocks [1, 2]. The CPT technology can significantly reduce the size, power and cost of atomic clocks [3-5]. Many CPT clocks use a vertical-cavity surface-emitting laser (VCSEL) [6] whose current is modulated by a generator with a frequency equal to half of the hyperfine splitting of the ground atomic state. Due to this modulation, the single-frequency laser mode acquires additional spectral components - sidebands. Usually two sidebands of the first order of the laser are used to excite CPT resonance in a two-photon λ scheme. Since the modulation frequency is half the hyperfine splitting of the ground-state atom, which is 6.8 GHz on the D1 line in ^{87}Rb, this is the case of modulation at 3.4 GHz. Another type of modulation considered here is the modulation of the laser current with a frequency equal to the hyperfine splitting of an atom in the ground state of 6.8 GHz. [7, 8]. With such modulation, the carrier and one of the first order sidebands are used to excite the CPT resonance according to the two-photon λ scheme. The CPT resonance can also be obtained at other laser current modulation frequencies that are multiples of the hyperfine splitting of the ground state of the atoms, but with increasing multiplicity, the spectrum of the side bands broadens, and the amplitudes of the two components participating in the two-photon λ-scheme become smaller. The rest of spectrum is produce background for the resonance by reducing contrast and increasing the optical noise.

The short-term stability of an atomic clock is inversely proportional to the resonance width and directly proportional to the signal-to-noise ratio determined by the resonance contrast and the shot noise of the radiation. Long term stability is determined by various frequency offsets. The dependence of the frequency shifts is investigated on operating parameters such as the light intensity, the value of the microwave power modulating the laser current, the optical detuning from the optical transition frequency, the temperature of the absorbing cell, the pressure of the buffer gas in the cell, and the external magnetic field.

This paper presents the results of studying the parameters of the CPT resonance in 87Rb on the D1 line and the light shift for the case of modulation of the laser current at frequencies of 3.4 GHz and 6.8 GHz.

2. Experimental setup

The setup (figure 1) includes a VCSEL operating at a wavelength of 795 nm, corresponding to the D1 line of ^{87}Rb. The laser is equipped with a precision current source and a temperature stabilization system. The laser output is linearly polarized and passes through the quarter-wave phase plate to obtain circular polarization. A gradient ND filter is used to modulate the radiation power level. It is slightly tilted relative to the position orthogonal to the beam to prevent reflections back into the laser. After that, the radiation is passed through an absorbing cell with rubidium vapor and a buffer gas. To observe absorption in alkali metal atoms, the cell is uniformly heated to more than 50 °C. Due to the design features, the cell windows are at a lower temperature. At the same time, metal condensation is observed on the windows. To solve this problem, the cell temperature T_c is maintained at 5 °C higher than the temperature of the T_f branch containing the metallic rubidium.

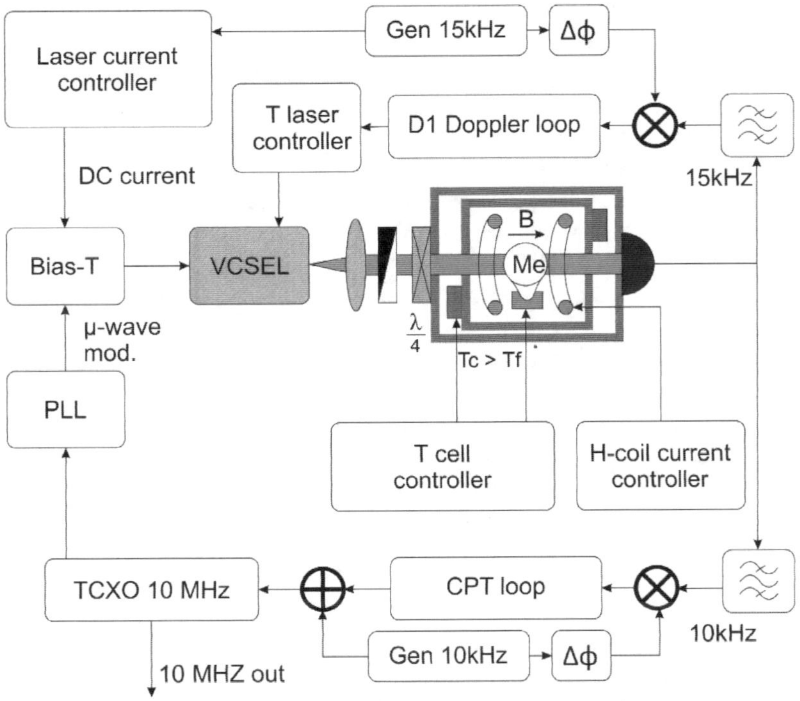

Figure 1. Schematic of an experimental setup for studying CPT resonances D1 of the ^{87}Rb line.

Helmholtz coils are located on the cell body to create a uniform magnetic field directed along the wave vectors of the waves. This field splits the energy levels of the ground state and allows one to observe the CPT resonance associated only with magnetic sublevels with quantum numbers $m_F = 0$, for which there is no linear Zeeman effect. At the same time, these levels undergo a quadratic shift in the magnetic field, which leads to a corresponding shift of the CPT resonance (frequency of the clock "0–0" transition):

$$\Delta\nu = K_0 B^2 , \qquad (1)$$

where B is the magnetic induction (in Gs). Based on the Breit-Rabi formula [9], it can be shown that for ^{87}Rb the proportionality coefficient is $K_0 = 575.14$ Hz/Gs^2. The magnitude of the magnetic field in our case was 100 - 150 mG.

Journal of Physics: Conference Series **2067** (2021) 012006 doi:10.1088/1742-6596/2067/1/012006

To obtain resonant optical frequencies for the λ scheme of two photon interaction ν_1 and ν_2 in the laser emission spectrum, its current was modulated at frequencies of 3.417 GHz and 6.834 GHz. For 3.4 GHz modulation, we used a proprietary microwave generator used as part of a miniature atomic clock, and for 6.8 GHz we used a Roghde & Schwarz SMA-100B RF and microwave signal generator. The frequencies ν_1 and ν_2 are side frequencies of orders ± 1 of the resulting frequency modulated (FM) laser radiation. In this case, the oscillator frequency is synthesized from a frequency of 10 MHz from a temperature compensated crystal oscillator (TCXO).

An automatic control system, operating at a laser current modulation frequency of 15 kHz, stabilizes the optical frequency of lasing along the D1 absorption line of ^{87}Rb. To obtain an error signal, the first harmonic of the modulation of the optical power of the beam passing through the rubidium cell is synchronously detected. The operating frequency of the probe modulation (15 kHz) was chosen on the basis that at this frequency, both the laser and the electronic elements usually do not have excessive flicker noise. You can control the laser generation frequency by changing its current or temperature. Current changes in speed are many times higher than the temperature method, but with a change in current, the output power of the laser, impedance, and modulation characteristics change to a much greater extent. Taking into account this feature, the processing of the error signal of the automatic control system is carried out by adjusting the laser temperature, and the test modulation through the laser current. Also, this approach compensates for the error of inaccuracy in measuring the temperature of the diode by the thermal sensor, associated with their different location in space.

To stabilize the frequency of the microwave generator, a stabilization method is used, which is similar to the Pound-Drever-Hall method used to stabilize the frequency of lasers in the optical range [10]. In this method, the modulation frequency significantly exceeds the resonance width. In our case, we use an automatic control system with a modulation frequency of ≈ 9 kHz with a full width at half maximum (FWHM) of the observed CPT resonance equal to 680 Hz. This method has a number of advantages, namely, it allows you to realize the maximum possible feedback speed, to select a high operating frequency to improve the signal-to-noise ratio.

The absorbing cell is made of quartz glass, resistant to alkali metals and having low gas permeability (for example, grade C51-1). The glass cell contains the pure isotope ^{87}Rb, a buffer inert gas (≈ 100 Torr). In the absence of a buffer gas, the time of coherent interaction of rubidium atoms with the laser field is much shorter and equal to the average time of flight of atoms through the beam, which leads to a significant broadening of CPT resonances.

3. Resonances with modulation of the laser current at 3.4 GHz frequency.

Sensitivity to the following parameters is the main source of long-term instability in manual CPT watches:

- intensity of optical radiation
- microwave power modulation
- collisional shift from buffer gases when the cell temperature changes
- shift from the change in the concentration of rubidium atoms with a change in the cell temperature
- shift from the magnetic field in the cell.

Changing the modulation frequency from 3.4 to 6.8 GHz affects only the first two shifts, the rest of the dependences will behave regardless of the modulation method [11].

The created experimental setup made it possible to observe CPT resonances, measure their parameters and, directly from the signal in the microwave frequency stabilization system, determine the frequency discriminator slope, both for the case of laser current modulation at a frequency of 3.4 GHz and at a frequency of 6.8 GHz. A typical record of the CPT resonance for the case of modulation at a frequency of 6.8 GHz is shown in figure 2, and the error signal in the microwave frequency stabilization system in figure 3.

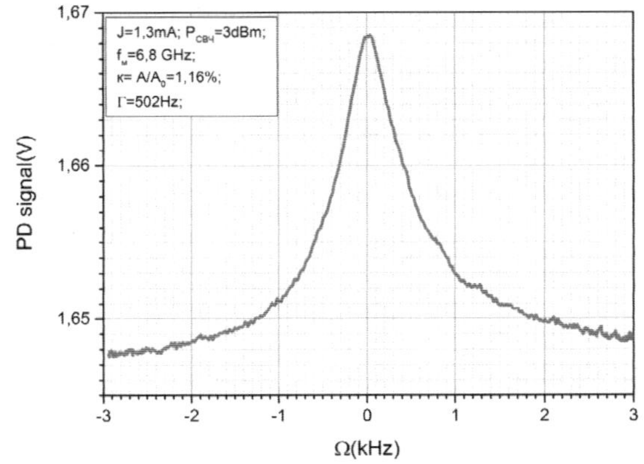

Figure 2. CPT-resonance.

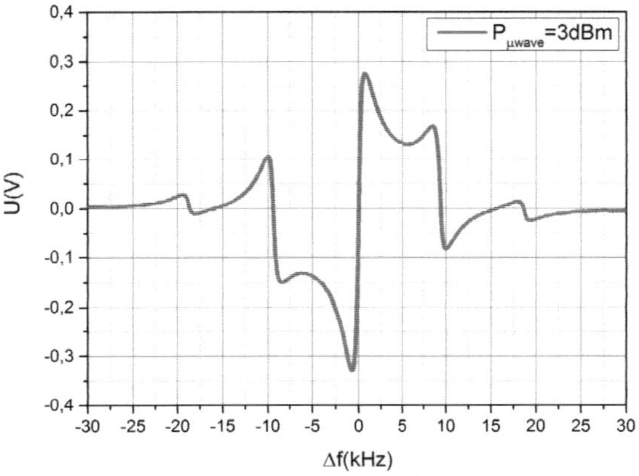

Figure 3. Error signal in the microwave frequency stabilization system.

To study the field frequency shifts as a function of the microwave power modulating the laser current, the microwave frequency was stabilized by the CPT resonance frequency and the frequency of the atomic standard was measured relative to the hydrogen standard frequency. For this, the 10 MHz output frequencies of the atomic standard based on ^{87}Rb and the Ch1-1007 hydrogen frequency and time standard were fed to the VCH-323 phase comparator - analyzer. The microwave power was changed stepwise at each step, and the frequency of the atomic standard was measured based on the CPT resonance in ^{87}Rb on the D1 line. The figure 4 shows the measurement results for the case of modulation of the laser current at a frequency of 3.4 GHz for two VCSEL.

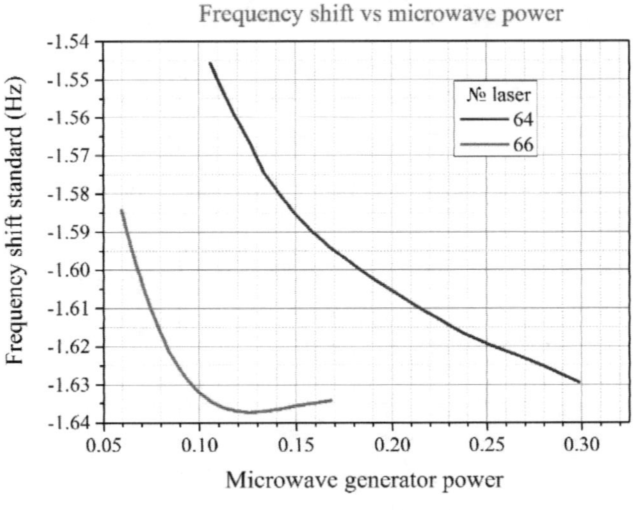

Figure 4. Frequency shift of the standard based on the CPT resonance in ^{87}Rb on the D1 line depending on the microwave power.

For one of the VCSEL, the dependence has an extremum, which we used earlier [11] to increase the long-term stability of the frequency of the atomic standard. At this point, the first derivative of the frequency shift versus microwave power is zero. But upon further investigation, it was found (figure 4) that such a point was not observed in another VCSEL series. To understand the difference between lasers, the spectra of lasers were studied using a scannable interferometer with a free dispersion region of ~40 GHz.

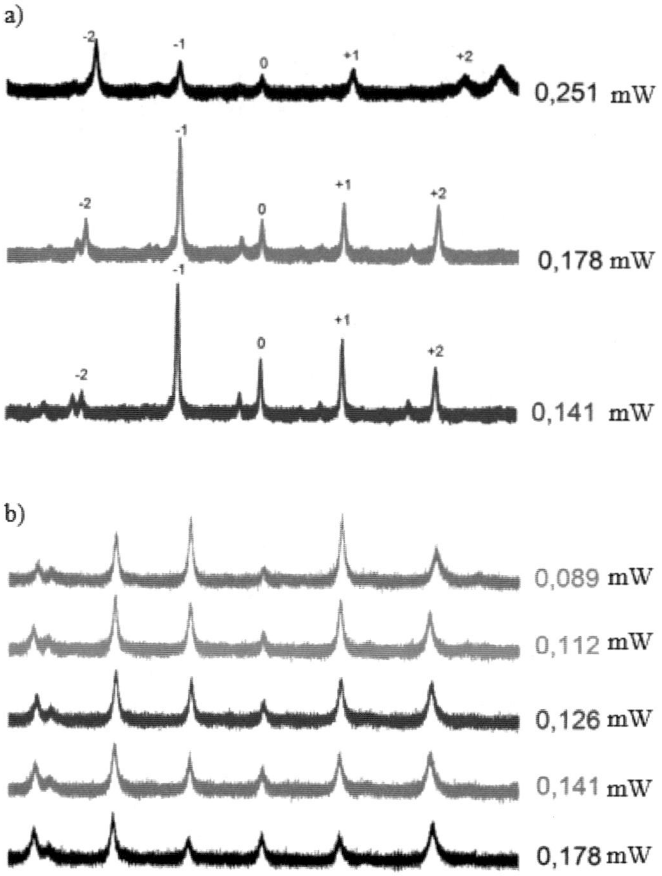

Figure 5. Spectra of lasers: a - №64, b - №66, at different levels of microwave modulation.

As shown in figure 5 spectra have sidebands located at intervals equal to the modulation frequency. In a laser for which a minimum was observed in the dependence of the frequency shift of the standard on the microwave power, the spectrum looks much more symmetric.

To find the optimal CPT stabilization parameters, the resonance was investigated by two main parameters: the slope of the frequency discriminator depending on the microwave power and the frequency shift of the atomic standard recalculated to 1% of the change in microwave power are plotted in figure 6 for the case of modulation of the laser current at a frequency of 3.4 GHz. The frequency shift of an atomic standard from a change in microwave power by 1% of the operating value of the microwave power is calculated through the first derivative at any selected operating point multiplied by 0.01 of the power value at this point (figure 6). The optimal point for long-term stability in microwave power is P = 0.16 mW, since with a small gap in the discriminator steepness, a zero shift from the microwave is observed, which makes the standard insensitive to such a poorly controlled parameter as the amplitude of the output signal of the microwave oscillator

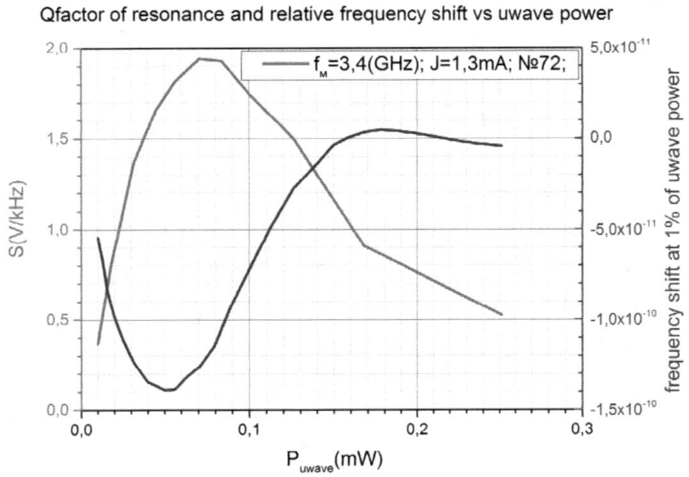

Figure 6. Measurements for the case of modulation of the laser current at a frequency of 3.4 GHz. S - dependence of the slope of the frequency discriminator on the microwave power. Frequency shift of an atomic standard when the microwave power is changed by 1%.

4. Resonances with modulation of the laser current at 6.8 GHz frequency.

Investigations were carried out in a similar way for the case of modulation of the VCSEL current at a frequency of 6.8 GHz. To begin with, a scanned interferometer was used to observe the laser spectra for the case of current modulation at a frequency of 6.8 GHz. The distribution of amplitudes in the spectral components is shown in figure 7. The side stripes are almost symmetrical.

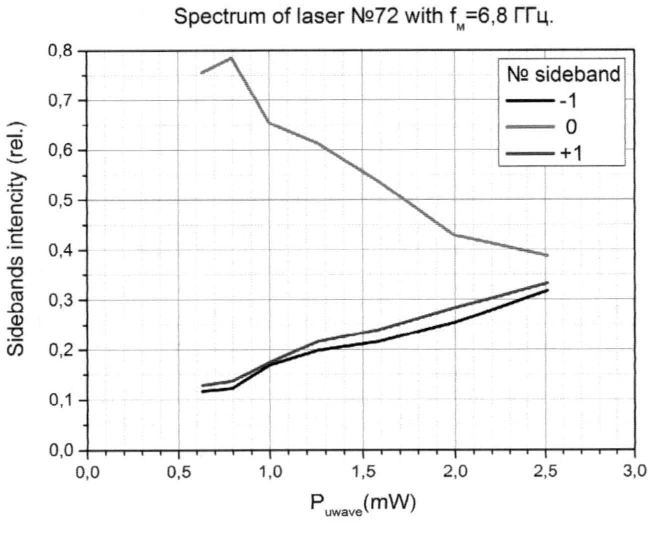

Figure 7. The amplitude of the sidebands of the laser with current modulation at a frequency of 6.8 GHz.

Investigations of CPT resonances and the slope of frequency discriminators were carried out for the case of modulation of the laser current at a frequency of 6.8 GHz and after processing all measurements, two main parameters are the slope of the frequency discriminator obtained on the basis of the CPT resonance depending on the microwave power and the frequency shift of the atomic standard recalculated to 1% of the change in microwave power plotted on one graph for the case of modulation of the laser current at a frequency of 6.8 GHz and are presented in figure 8. A zero shift depending on the microwave power was not observed for this type of modulation; therefore, the optimal laser operation mode was selected according to the maximum discriminator slope at a microwave power of 2.5 mW.

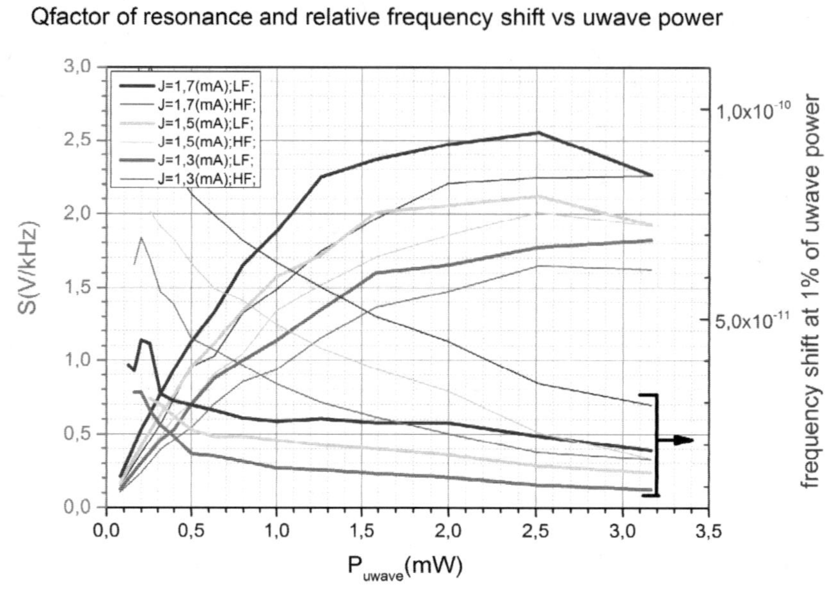

Figure 8. Measurements for the case of modulation of the laser current at a frequency of 6.8 GHz.

Dependence of the slope of the frequency discriminator on the microwave power. Frequency shift of an atomic standard when the microwave power is changed by 1%.

For the case of modulation of the LWR current at a frequency of 3.4 GHz, measurements were carried out only for the laser current J = 1.3 mA. For the case of modulation of the LWR current at a frequency of 6.8 GHz, measurements were performed for three values of the laser current J = 1.7 mA, J = 1.5 mA, and J = 1.3 mA. In addition, the λ scheme for observing the CPT resonance for the case of modulation of the LVD current at a frequency of 6.8 GHz can be constructed in two versions using the low-frequency sideband and the central lasing band of the laser, and the second version using the high-frequency sideband and the central lasing band of the laser. In total, there are six different options.

Further, the field frequency shifts of the standard were investigated for the case of a modulation frequency of 6.8 GHz. The measurement results for all six cases are presented in figure 9.

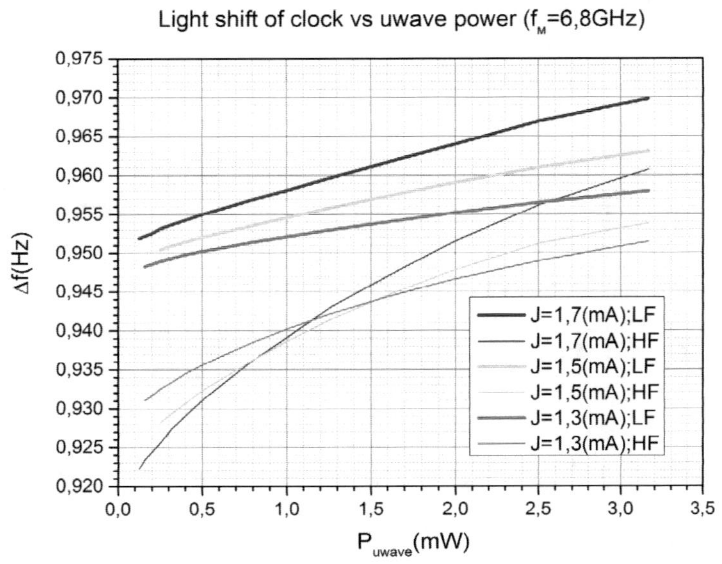

Figure 9. Field frequency shifts of the standard for the case of a modulation frequency of 6.8 GHz depending on the microwave power for various laser currents. CPT resonance was obtained from different components of the radiation spectrum: LF - carrier + low-frequency sideband, HF - carrier + low-frequency sideband.

It can be seen from the graphs that the shifts from the microwave modulation power are significantly less when stabilized behind the absorption line by a combination of carrier + low-frequency side bands (LF), and when using the HF mode, a point of low sensitivity to changes in the laser current is observed. At the same time, the stability of the current source is a fairly well controlled parameter; therefore, the LF resonance observation mode was chosen as optimal.

After choosing the optimal stabilization parameters, an experiment was set up to measure the Alan function for both cases of modulation (figure 10). It can be seen from the graphs that for 6.8 GHz, a significant improvement was obtained in short times $\sim 3 \cdot 10^{-12}\ \tau^{-1/2}$. The instability over long periods was due to the relatively high temperature sensitivity of the Roghde & Schwarz SMA-100B microwave generator.

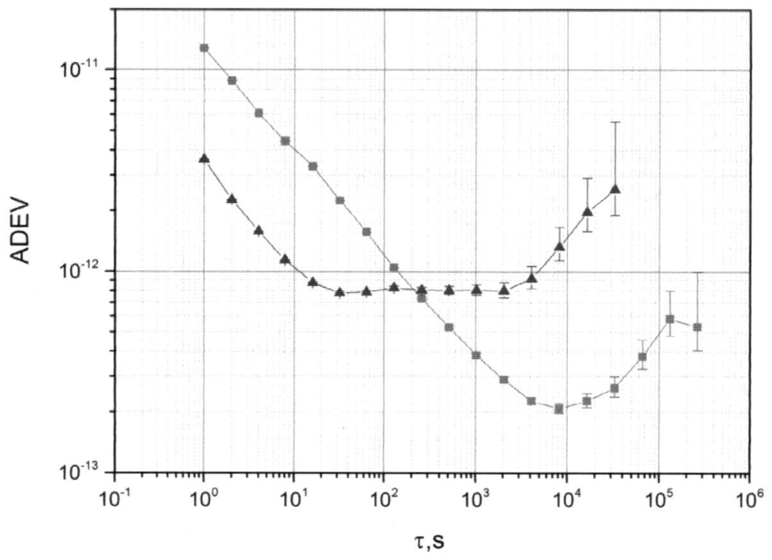

Figure 10. Results of measuring the Alan function for the case of modulation of the laser current at frequencies of 3.4 (red squares) and 6.8 GHz (blue triangles).

5. Conclusion

The studies carried out comparing the parameters of the CPT resonance in 87Rb on the D1 line for the cases of modulation of the laser current at frequencies of 3.4 GHz and 6.8 GHz showed that the slope of the frequency discriminator obtained on the basis of the CPT resonance at the optimal microwave power is the same in both cases. Studies of the field shifts of the frequency standard based on CPT resonances have shown that in the case of modulation of the laser current at a frequency of 6.8 GHz, the field shifts are 2–4 times smaller. The achieved result on short-term stability of 3 · 10-12 for 1 s showed that this approach is promising for use in miniature atomic clocks.

Acknowledgments

Our team would like to thank the Russian Foundation for Basic Research for the support of the work by grant 20-32-90029.

References

[1] Knappe S et al., 2001 Characterization of coherent population-trapping resonances as atomic frequency references, *JOSA B.*, **18**, pp 1545-1553
[2] Letokhov V S and Chebotayev V P 1977 Nonlinear laser spectroscopy *Springer, Berlin*
[3] Kitching J 2018 Chip-scale atomic devices *Applied Physics Review*, **5** 031302 –031302-38
[4] Knappe S, Schwindt P D D, Shah V et al. 2005 A chip-scale atomic clock based on 87Rb with improved frequency stability, *Optics Express*, **13** 1249-1253
[5] Knappe S et al. 2004 A microfabricated atomic clock *Applied Physics Letters* **85** 1460-1462
[6] Michalzik R 2013 VCSEL fundamentals, VCSELs, *Springer, Berlin, Heidelberg*, pp 19-75
[7] Goka S, Okura T, Moroyama M and Watanabe Y 2008 Low Power 85Rb CPT Atomic Clock *IEEE International Frequency Control Symposium*
[8] Deng K, Guo T, Su J, Guo D, Liu X, Liu L, Chen X, Wang Z 2009 Full hyperfine frequency

modulation in the implementation of coherent population trapping atomic clocks, *Physics Letters A* **373** 1130–1132

[9] Miletic D 2013 Light-shift and temperature-shift studies in atomic clocks based on coherent population trapping, *Université de Neuchâtel.*

[10] Drever R W P et al. 1983 Laser phase and frequency stabilization using an optical resonator, *Applied Physics B*, **31**, pp 97-105

[11] Skvortsov M N et al. 2020 Miniature quantum frequency standard based on the phenomenon of coherent population trapping in vapours of 87Rb atoms, *Quantum Electronics*, **50**, pp 576

Journal of Physics: Conference Series **2067** (2021) 012007 doi:10.1088/1742-6596/2067/1/012007

Spectroscopic properties of CrO_x/Al_2O_3 nanopowders synthesized by cw CO_2 laser vaporization

M G Baronskiy*, A I Kostyukov and V N Snytnikov

Boreskov Institute of Catalysis SB RAS, Lavrentiev Ave. 5, Novosibirsk 630090, Russia

E-mail: baronskiymg@mail.ru

Abstract. Nanosized 5.0 wt% Cr/nano-Al_2O_3 powders with the particle size of ca. 15 nm were synthesized via laser vaporization using irradiation by a cw CO_2 laser in different gas atmospheres – Ar, Ar+O_2, Ar+H_2. All the investigated nanopowders were studied by XRF, XRD, TEM, UV-vis DRS and PL spectroscopy methods. The nanopowders were found to contain the Cr^{6+} ions located on the surface of Al_2O_3 nanoparticles and two types of Cr^{3+} sites. One type is Cr^{3+} ions located in the bulk ($Cr^{3+}{}_b$-sites) of Al_2O_3 matrix in a strong crystal field. The second type is represented by Cr^{3+} sites residing near the surface ($Cr^{3+}{}_s$-sites) of CrO_x/Al_2O_3 nanoparticles in a weak crystal field. It was shown that varying the composition of the buffer gas (Ar, Ar+O_2, Ar+H_2) during laser vaporization makes it possible to control the properties of the obtained 5.0 wt% Cr/nano-Al_2O_3 nanopowders with a change in Cr^{6+}/Cr^{3+} ratio in the bulk and on the surface of alumina support.

1. Introduction

Today, the development of methods for the synthesis of nanoscale materials has greatly increased interest to them, since the transition from coarse-crystalline materials to nanoscale analogs is often accompanied by a significant change in the physicochemical properties of substances. It was found that the developed surface of individual nanocrystallites affects their bulk structure and thereby contributes to their properties [1, 2]. One of the leading approaches in the synthesis of nanomaterials is the "bottom-up" approach. It consists of the stepwise formation of nanomaterials according to the sequence "atoms → clusters → nanoparticles" [2]. A striking example of this approach is laser synthesis methods, which are based on the Physical Vapor Deposition (PVD) method. Laser methods for the synthesis of oxide nanoparticles are presented mainly by the methods of laser vaporization in the gas phase (LAVA and co-LAVA) [3, 4] and pulsed laser ablation (PLA) in a liquid [5, 6]. Laser synthesis methods make it possible to produce nanoparticles by evaporation of starting materials by exposure to laser radiation and subsequent condensation of the vapor of the evaporated substance in a gas or liquid, the so-called buffer gas or liquid [7-10]. By varying several parameters in the process of laser evaporation such as laser radiation power, the composition of the buffer gas in the evaporation chamber, and gas pressure it becomes possible to control the physicochemical properties of the resulting nanoparticles. Thus, one of the main advantages of laser synthesis methods is high purity, in contrast to, for example, chemical methods, and monodispersity of the obtained samples, as well as the ability to control the size of nanoparticles in the range from 1 to tens of nanometers [11-13]. Laser synthesis methods are actively used to obtain simple oxides, for example, Al_2O_3, ZrO_2, TiO_2, Y_2O_3, SiO_2, as well as multicomponent compounds such as YSZ, YAG, Eu:$SrAl_2O_4$, etc. [12, 14-16].

In the present work, we investigated the physicochemical properties of the nanosized CrO_x/Al_2O_3 systems synthesized by cw CO_2 laser vaporization in different gas atmospheres – Ar, O_2, H_2. The samples under consideration are of interest as promising heterogeneous catalysts for isobutane dehydrogenation reaction. A characteristic feature of CrO_x/Al_2O_3 catalysts is a wide variety of different species of supported chromium oxide particles [17, 18]. This is largely due to the peculiarities of the catalyst preparation, as well as the significant influence of the chemical nature of the initial support on the properties of the final deposited chromium oxides on Al_2O_3 surface. This influence manifests itself in a strong interaction of the supported chromium oxide with the alumina support surface, as a result of which chromium ions are stabilized in different oxidation states and different coordination. There are no data in the literature on the study of the formation of various chromium sites in the nanosized CrO_x/Al_2O_3 systems prepared by laser evaporation, except for the work of the authors of this paper [19]. The use of various gas atmospheres (oxidative and reductive) during laser synthesis of CrO_x/Al_2O_3 catalysts should lead to the emergence of different charge states of chromium, a change in their ratio, which, as a result, will make it possible to synthesize chromia-alumina catalysts with a higher concentration of the active component.

The main aim of the work is the study of various charge states of chromium in 5.0wt% Cr/nano-Al_2O_3 powders, synthesized by cw CO_2 laser vaporization. Establishing the effect of the atmosphere (Ar, O_2, H_2) during vaporization on the ratio of the main types of charge states of chromium Cr^{3+}/Cr^{6+} in the investigated nanopowders. This paper is a consistent continuation of our work for the study of different Cr^{n+} species in chromia-alumina systems prepared by different methods: zol-gel [20], laser synthesis [19].

2. Experimental

2.1. Preparation of samples

The Cr:α-Al_2O_3 ceramic targets for further laser vaporization as pellets (diameter 18 mm, thickness 10 mm, density of the targets 1.8 g/cm^2) were prepared using a highly dispersed γ-Al_2O_3 powder (99%) as a starting material. γ-Al_2O_3 was obtained by calcination of pseudoboehmite (γ-AlOOH×0.37H_2O) for 4 h at a temperature of 550°C. To increase the Cr concentration in the targets, the γ-Al_2O_3 powder before pelletization was modified using chromium nitrate $Cr(NO_3)_3\times 9H_2O$ (99.7%) via incipient wetness impregnation with an aqueous solution of nitrate salt with chromium concentration 5.0 wt%. The Cr:γ-Al_2O_3 powder was then loaded in a vacuum press mold and pelletized at a force of ca. 13–15 t. After that, the Cr:γ-Al_2O_3 pellets were calcined in a crucible at 1250 °C for 4 h until Cr:α-Al_2O_3 was formed. According to XRF data, the Cr concentration in Cr:α-Al_2O_3 targets was 4.98±0.04 wt%.

5.0 wt% Cr/nano-Al_2O_3 nanopowders were obtained by laser vaporization of Cr:α-Al_2O_3 ceramic targets irradiated by a cw CO_2 laser (radiation wavelength 10.6 μm, generation power up to 110 W on one TEM_{00} transverse mode, output beam diameter 8 mm, divergence in the far-field region 3×10^{-3} rad) with subsequent condensation of vapor in a buffer gas flow in a vaporization chamber. Samples of nanopowders were synthesized in argon (99.998%) and argon supplemented with oxygen (99.7%) and hydrogen (99.999%) in the concentration 20 and 30 vol%, respectively. Laser power on the target surface was 103 W (the power density 5.5×10^4 W/cm^2). After exposure of the target surface to laser radiation, it was heated and the target material was evaporated with subsequent vapor condensation in the low-temperature zone of the chamber. A gas-dust flow with 5.0 wt% Cr/nano-Al_2O_3 particles was then passed through the filter where the nanoparticles were settled and then collected for further investigation. During all the procedure of laser synthesis, Ar pressure in the vaporization chamber was 0.1 atm. A detailed description of the experimental setup used for the laser vaporization of nanomaterials with different chemical composition can be found in Refs. [12, 21].

2.2. Characterization of samples

The chemical composition of the investigated samples was controlled by X-ray fluorescence (XRF) analysis on an ARL – Advant'x analyzer.

X-ray diffraction (XRD) and transmission electron microscopy (TEM) data were obtained for 5.0 wt% Cr/nano-Al_2O_3 nanopowder synthesized in Ar atmosphere. XRD pattern was obtained on a Bruker D8 Advance diffractometer using CuKα radiation (λ = 0.15418 nm). Measurements were performed in the 2θ range of 10–70° with a step 0.05° and acquisition time 3 s. Phases were identified by comparing experimental diffraction patterns with the data of ICDD, PDF 2 database.

TEM images were collected on a JEM-2010 electron microscope at accelerating voltage 200 kV and resolution 1.4 Å. During TEM measurements all the samples were deposited on a copper grid by dispersing a solid phase suspension in alcohol using an ultrasonic disperser.

Photoluminescence (PL) and photoluminescence excitation (PLE) spectra were measured on a CaryEclipse (Varian) fluorescence spectrophotometer with a Xe lamp as an excitation source. UV-vis DRS measurements were carried out on a UV 2501 PC (Shimadzu) spectrophotometer with an ISR 240A diffuse reflectance attachment. The nanopowders under consideration for PL, PLE, and UV-vis DRS experiments were placed in quartz cuvettes.

3. Results and Discussion

According to XRF data, the Cr concentration in studied nanopowders was 4.8±0.05 wt%. Figure 1 demonstrates the photographs of the initial Cr:α-Al_2O_3 ceramic target and final 5.0 wt% Cr/nano-Al_2O_3 nanopowders synthesized in different gas atmospheres – Ar, Ar+O_2, Ar+H_2. As seen in Fig. 1, depending on the composition of the atmosphere during vaporization, all the resulting powders have a different color.

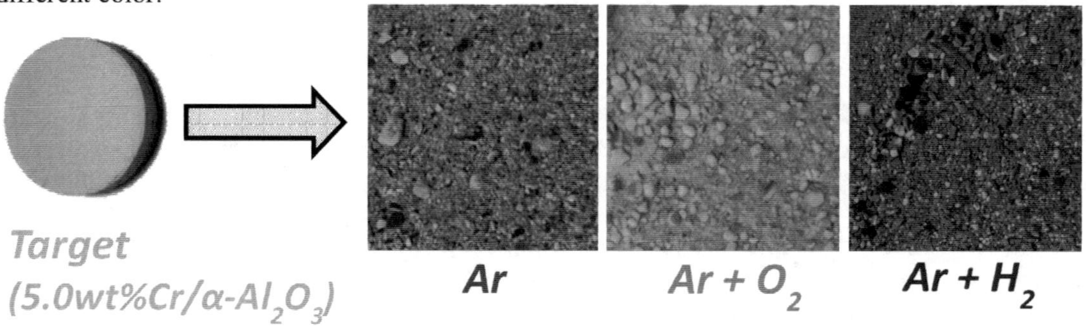

Figure 1. Photographs of Cr:α-Al_2O_3 ceramic target and laser vaporized 5.0 wt% Cr/nano-Al_2O_3 nanopowders.

Figure 2 displays XRD pattern and TEM images of 5.0 wt% Cr/nano-Al_2O_3 nanopowder synthesized in Ar. X-ray diffraction pattern demonstrates that in terms of phase composition the sample corresponds predominantly to γ-Al_2O_3 with the beginning of the transition to δ-Al_2O_3.

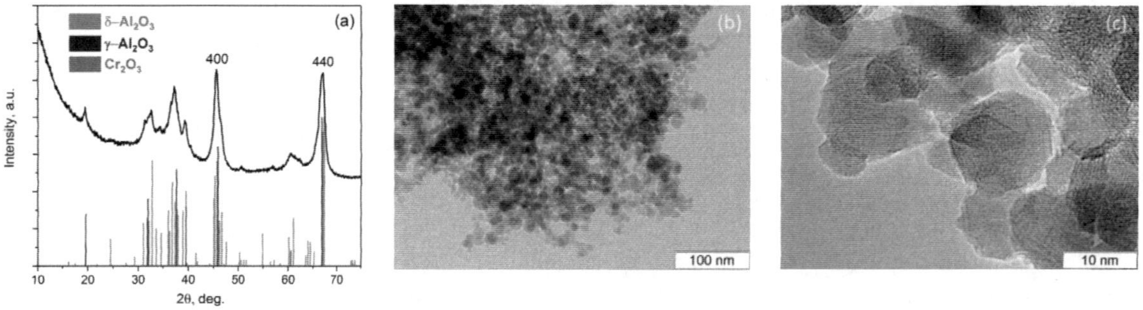

Figure 2. XRD pattern (a) and TEM images (b, c) of 5.0 wt% Cr/nano-Al_2O_3 nanopowder synthesized in Ar.

TEM images demonstrate that the studied sample is represented by faceted spherically symmetric nanoparticles with an average size of d_m= 15 nm. It should be noted that the XRD and TEM data for other nanopowders studied in this work are similar to the data for 5.0 wt% Cr/nano-Al_2O_3 synthesized in Ar.

Figure 3 shows UV-vis DRS, PL, and PLE spectra of the investigated 5.0 wt% Cr/nano-Al_2O_3 nanopowders. The obtained UV-vis DRS spectra demonstrate the presence of four bands with maxima at 16500, 22500, 27100, and 36600 cm^{-1} in all studied samples. All the PL spectra in each case demonstrate one broad band with a maximum at ~14300 cm^{-1}. PLE spectra contain two broad bands with maxima at ~17700 and 23400 cm^{-1}, respectively.

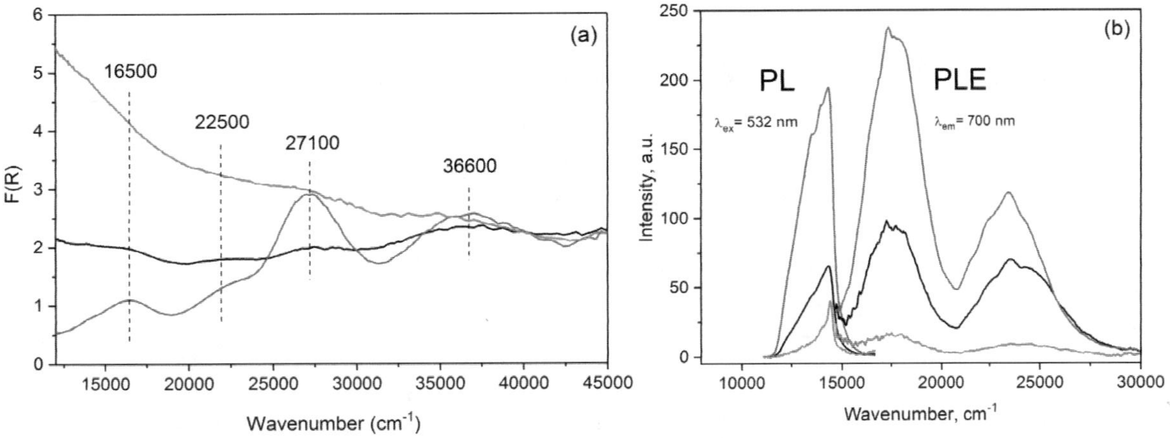

Figure 3. UV-vis DRS spectra (a) and PL, PLE spectra (b) of 5%Cr/nano-Al_2O_3 nanopowders synthesized in different gas atmospheres during vaporization: black curve – Ar; pink curve – Ar + H_2; green curve – Ar + O_2. PL/PLE: λ_{ex}= 532 nm (18800 cm^{-1})/λ_{em}= 700 nm (14400 cm^{-1}).

Based on the analysis of spectroscopic data, it was concluded that in the studied 5.0 wt% Cr/nano-Al_2O_3 nanopowders, chromium is stabilized mainly in the charge states of Cr^{6+} and Cr^{3+} in different coordination. Thus, according to the UV-vis DRS data (see Fig.3 a), the absorption bands at 16500 and 22500 cm^{-1} correspond to *d-d* transitions in Cr^{3+} ions located in octahedral oxygen coordination in the matrix of nanosized Al_2O_3 ($^4A_{2g}\rightarrow{}^4T_{2g}$ and $^4A_{2g}\rightarrow{}^4T_{1g}$ electron transitions, respectively) [22, 23]. The bands at 27100 and 36600 cm^{-1} in UV-vis DRS spectra correspond to the ligand-metal CT bands for Cr^{6+} ions in tetrahedral coordination of oxygen ions [23]. An increase in the intensity and a change in the ratio of the corresponding bands at 27100 and 36600 cm^{-1} in the UV-vis DRS spectra for the sample obtained by laser vaporization in the presence of O_2 indicates an increase in the concentration of water-soluble forms of Cr^{6+} in tetrahedral coordination. The variety of different chromium compounds, the defectiveness of the structure of the obtained nanopowders, most likely, is responsible for their differences in color.

Chromium ions in the trivalent state, that are located in the bulk of alumina matrix, exhibit intense luminescence in the red spectral region [24-27]. Thus, the band at ~14300 cm^{-1} in the PL spectra (Fig. 3 b) is identified with the radiative *d-d* transition $^2E_g\rightarrow{}^4A_{2g}$ in octahedrally coordinated Cr^{3+} ions in Al_2O_3 lattice. The broad bands in the PLE spectra correspond to the $^4T_{2g}$, $^4T_{1g}\rightarrow{}^4A_{2g}$ electron transitions for Cr^{3+} ions in octahedral coordination [19].

Earlier, in xCr/Al_2O_3 (zol-gel synthesis; [x]= 0.25, 0.5, 1.0 wt%) and xCr/nano-Al_2O_3 (laser synthesis; [x]= 0.0, 0.05, 0.5, 1.0, 2.5, 5.0 wt%) samples, along with the emission of Cr^{3+} ions ($^2E_g\rightarrow{}^4A_{2g}$ electron transition) located in the bulk of Al_2O_3 lattice, we revealed the PL of Cr^{3+} sites located in more disordered structural positions, which are situated in the subsurface layers of Al_2O_3 nanocrystallites [19, 20]. This PL corresponds to the $^4T_2\rightarrow{}^4A_2$ electron transition in Cr^{3+} ions, which are in a weak crystal field, where Dq/B≤ 2. Electron transition $^4T_2\rightarrow{}^4A_2$ becomes possible when the octahedron in which the Cr^{3+} ion is located undergoes significant distortion, leading to the position of the $^4T_2$ state on the Tanabe-Sugano energy diagram appears at lower energies than the ^2E state [28, 29]. Figure 4 shows the PL spectra (λ_{ex}= 532 nm (18800 cm^{-1})) of 5.0 wt% Cr/nano-Al_2O_3 nanopowders deconvolved into Gaussian components.

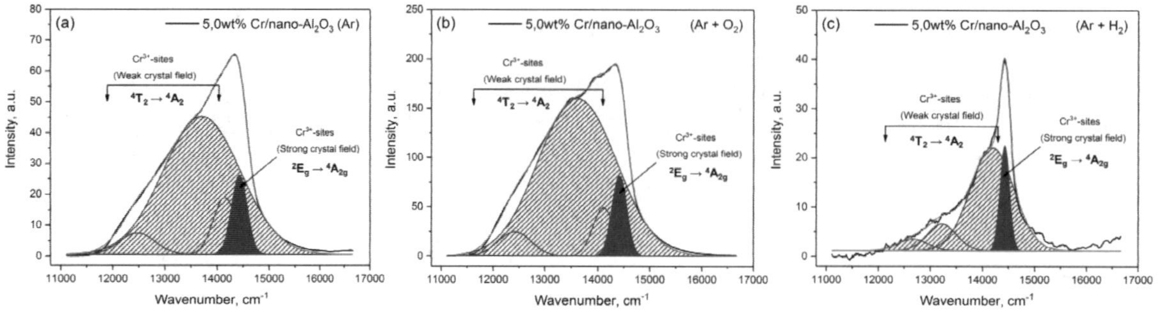

Figure 4. PL spectra of 5.0 wt% Cr/nano-Al_2O_3 nanopowders synthesized in different gas atmospheres – Ar, Ar + O_2, Ar + H_2. λ_{ex}= 532 nm (18800 cm^{-1}) (with deconvolution into Gaussians).

Deconvolution of the PL spectra into Gaussian components made it possible to reveal two types of Cr^{3+} sites in all the studied samples. The first type is Cr^{3+} ions located in the bulk ($Cr^{3+}{}_b$) of Al_2O_3 matrix, in a strong crystal field. $Cr^{3+}{}_b$ sites correspond to the PL band with λ_{max}~ 14400 cm^{-1} ($^2E_g \rightarrow {}^4A_{2g}$ electron transition). Whereas Cr^{3+} ions in nonequivalent positions corresponding to a weak crystal field ($^4T_2 \rightarrow {}^4A_2$ electron transition) are responsible for the shoulders with λ_{max}~ 12500, 13600, 14100 cm^{-1} for 5.0 wt% Cr/nano-Al_2O_3 (Ar, Ar+O_2) and 12700, 13300, and 14200 cm^{-1} for 5.0 wt% Cr/nano-Al_2O_3 (Ar+H_2). These are Cr^{3+} sites that are located in the subsurface ($Cr^{3+}{}_s$) layers of Al_2O_3 nanocrystallites.

4. Conclusion

The nanostructured powders Cr/nano-Al_2O_3 with the Cr concentration 5.0 wt% were synthesized via laser vaporization using a radiation by a cw CO_2 laser in flowing argon and in Ar with the addition of O_2 and H_2. Laser synthesized Cr/nano-Al_2O_3 nanopowders can be used as the promising catalysts for isobutane dehydrogenation.

As follows from TEM images investigated nanopowders are represented by faceted spherically symmetric nanoparticles with an average size of d_m= 15 nm. XRD data demonstrate that in terms of phase composition the samples correspond predominantly to low-temperature γ-Al_2O_3 with the beginning of the transition to high-temperature δ-Al_2O_3. The electronic states of various chromium ions were investigated by means of UV-vis DRS and PL spectroscopy methods. It was shown that Cr ions in the 5.0 wt% Cr/nano-Al_2O_3 samples are stabilized predominantly in Cr^{3+} and Cr^{6+} states in octahedral and tetrahedral coordination, respectively. The PL properties of all the studied nanopowders are caused by the luminescence of octahedrally coordinated Cr^{3+} ions in Al_2O_3 matrix. The analysis of the acquired data allowed us to separate the observed PL into luminescence of Cr^{3+} sites located in the bulk of Al_2O_3 lattice ($Cr^{3+}{}_b$-sites; the case of a strong crystal field) and Cr^{3+} sites located near the surface of Al_2O_3 nanocrystallites ($Cr^{3+}{}_s$-sites; the case of a weak crystal field).

Varying the composition of the buffer gas (Ar, Ar+O_2, Ar+H_2) during laser vaporization makes it possible to control the properties of the obtained 5.0 wt% Cr/nano-Al_2O_3 nanopowders with a change in Cr^{6+}/Cr^{3+} ratio in the bulk and on the surface. This is particularly important in catalytic studies on the dehydrogenation of alkanes using laser vaporized CrO_x/Al_2O_3 nanosized powders as novel catalytic systems.

Acknowledgements
The authors are grateful to Vl.N. Snytnikov for laser synthesis of nanopowders, to PhD A.A. Zhdanov for elemental analysis by XRF method, to PhD S.V. Cherepanova for XRD data, to PhD A.V.

Ishchenko for TEM images of studied nanopowders, and to PhD T.V. Larina for UV-vis DRS measurements.

The study was supported by the Russian Foundation for Basic Research within project No. 20-43-540005 and partially by the Ministry of Science and Higher Education of the Russian Federation within the governmental order for Boreskov Institute of Catalysis SB RAS (project AAAA-A21-121011390009-1).

References

[1] Rempel A A 2007 *Russ. Chem. Rev.* **76** (5) 435-461.
[2] Gangwar J, Gupta B K, Tripathi S K, Srivastava A K 2015 *Nanoscale* **32**(7) 13313-13344.
[3] Kurland H-D, Grabow J, Muller F A 2011 *Journal of the European Ceramic Society* **31** 2559-2568.
[4] Bartolomé J F, Smirnov A, Kurland H-D, Grabow J, Müller F A 2016 *Scientific Reports* **6** 20589:1-11.
[5] Sajti C L, Sattari R, Chichkov B, Barcikowski S 2010 *Applied Physics A* **100** 203-206.
[6] Lam J, Amans D, Chaput F, Diouf M, Ledoux G, Mary N, Masenelli-Varlot K, Motto-Rosa V, Dujardin C 2014 *Phys.Chem.Chem.Phys* **16** 963-973.
[7] Kim M, Osone S, Kim T, Higashi H, Takafumi S 2017 *KONA Powder and Particle Journal* **34** 80-90.
[8] Kostyukov A I, Snytnikov Vl N, Zhuzhgov A V, Cherepanova S V, Ishchenko A V, Baronskiy M G, Snytnikov V N 2020 *Journal of Alloys and Compounds* **815** 152476:1-12.
[9] Kostyukov A I, Snytnikov V N, Snytnikov Vl N, Ishchenko A V, Rakhmanova M I, Molokeev M S, Krylov A S, Aleksandrovsky A S 2020 *Optical Materials* **104** 109843:1-5.
[10] Kostyukov A I, Snytnikov Vl N, Snytnikov V N, Rakhmanova M I, Kostyukova N Y, Ishchenko A V, Cherepanova S V, Krylov A S, Aleksandrovsky A S 2021 *Journal of Luminescence* **235** 118050:1-11.
[11] Yang X C, Riehemann W, Dubiel M, Hofmeister H 2002 *Materials Science and Engineering B* **95** 299-307.
[12] Snytnikov V N, Snytnikov Vl N, Dubov D A, Zaikovskii V I, Ivanova A S, Stoyanovskii V O, Parmon V N 2007 *Journal of Applied Mechanics and Technical Physics* **48**(2) 292-302.
[13] Kostyukov A I, Snytnikov Vl N, Yelisseyev A P, Zhuzhgov A V, Kostyukova N Y, Ishchenko A V, Cherepanova S V, Snytnikov V N 2021 *Advanced Powder Technology* **32** 2733-2742.
[14] Guo D, Ito A, Goto T, Tu R, Wang C, Shen Q, Zhang L 2013 *J. Adv. Ceram.* **2** 162-166.
[15] Zollfrank C, Gruber S, Batentschuk M, Osvet A, Goetz-Neunhoeffer F, Dittrich S, Grabow J, Kurland H-D, Muller F A 2013 *Acta Materialia* **61** 7133-7141.
[16] Kimura T, Goto T 2003 *Materials Transactions* **44**(3) 421-424.
[17] Puurunen R, Weckhuysen B 2002 *Journal of catalysis* **210** 418-430.
[18] Gaspar A, Brito J, Dieguez L 2003 *J. Mol. Catal. A* **203** 251-266.
[19] Kostyukov A I, Baronskiy M G, Larina T V, Snytnikov Vl N, Zaitseva N A, Pochtar A A, Ishchenko A V, Cherepanova S V, Snytnikov V N 2020 *Materials Characterization* **169** 110664:1-14.
[20] Baronskiy M G, Kostyukov A I, Larina T V, Snytnikov V N, Zaitseva N A, Zhuzhgov A V 2019 *Materials Chemistry and Physics* **234** 403-410.
[21] Kostyukov A I, Snytnikov V N, Rakhmanova M I, Kostyukova N Y, Snytnikov V N 2020 *Optical Materials* **110** 110508:1-6.
[22] Lever A 1987 *Inorganic Electronic Spectroscopy*, Elsevier, Amsterdam, Oxford, New-York, Tokyo, 863.
[23] Weckhuysen B, Schoonheydt R 1999 *Catalysis Today* **49** 441-451.
[24] Rastorguev A A, Baronskiy M G, Zaitseva N A, Isupova L A, Kostyukov A I, Larina T V, Pakhomov N A, Snytnikov V N 2014 *Inorganic Materials: Applied Research* **5** (5) 476-481.
[25] Rastorguev A, Baronskiy M, Zhuzhgov A, Kostyukov A, Krivoruchko O, Snytnikov V 2015

RSC Advances **5** 5686-5694.
[26] Baronskiy M, Rastorguev A, Zhuzhgov A, Kostyukov A, Krivoruchko O, Snytnikov V 2016 *Optical Materials* **53** 87-93.
[27] Kostyukov A, Baronskiy M, Rastorguev A, Snytnikov V, Snytnikov Vl, Zhuzhgov A, Ishchenko A 2016 *RSC Advances* **6** (3) 2072-2078.
[28] Pillonnet A, Garapon C, Champeaux C, Bovier C, Jaffrezic H, Mugnier J 2000 *Journal of Luminescence* **87-89** 1087-1089.
[29] Brik M G, Camardello S J, Srivastava A M, Avram N M, Suchocki A 2016 *ECS Journal of Solid State Science and Technology* **5** (1) R3067- R3077.

Luminescence performance of laser synthesized $Al_2O_3:Eu^{3+}$ nanophosphors depending on synthesis conditions

A Kostyukov[1, 2], A Nashivochnikov[1, 2], M Rakhmanova[3], and V Snytnikov[2]

[1] Novosibirsk State University, Pirogova Str. 2, 630090, Novosibirsk, Russia

[2] Boreskov Institute of Catalysis SB RAS, 5 Lavrentiev Ave., 630090, Novosibirsk, Russia

[3] Nikolaev Institute of Inorganic Chemistry SB RAS, 3 Lavrentiev Ave., 630090, Novosibirsk, Russia

Corresponding author's e-mail: ant.kostyukov@mail.ru (A. I. Kostyukov)

Abstract. $Al_2O_3:Eu^{3+}$ nanophosphors were prepared by laser vaporization method in a flowing mixture of Ar/He and O_2. Luminescence properties of $Al_2O_3:Eu^{3+}$ nanophosphors are predominantly determined by Eu^{3+} ions red emission with inhomogeneously broadened bands in the region of 550-750 nm corresponding to $^5D_0 \rightarrow {}^7F_J$ transitions of Eu^{3+} ($J = 0 - 4$). Hypersensitive electric dipole transition $^5D_0 \rightarrow {}^7F_2$ dominates in the spectrum and is responsible for the red emission. The effect of crystallite size on luminescence properties of $Al_2O_3:Eu^{3+}$ nanocrystals was observed. It was shown that the introduction of oxygen during the synthesis improves the luminescence performance. The obtained chromaticity coordinates and high absolute QY (~ 14%) indicate the possibility of using red nanophosphors based on $Al_2O_3:Eu^{3+}$.

1. Introduction

Currently, investigations of the nanophosphors with red emission are of interest to the material research community [1-8]. Alumina (Al_2O_3) is one of the standard phosphor matrices for activators emitting in the red spectral region [6, 9, 10]. Al_2O_3 possesses some unique properties, among which is a large transparency window from the short UV to the near-IR spectral region, and has good mechanical properties and excellent chemical stability. Therefore, Al_2O_3 is a suitable host for Eu^{3+} ions. Eu^{3+} ions in Al_2O_3 combine an intense red photoluminescence and a relatively easily interpretable luminescence spectrum. There are low-temperature η-, γ-, χ-, and high-temperature δ-, κ-, θ-, α- polymorphs of Al_2O_3. The only thermodynamically stable phase of Al_2O_3 is α polymorph (corundum). All metastable phases of Al_2O_3 irreversibly transform into the α- polymorph upon heat treatment over 1200 °C of the starting hydroxides and oxides. The photoluminescent properties of Eu^{3+} ions are best studied in α-Al_2O_3 and γ-Al_2O_3 matrices [11, 12]. When studying nano-sized Al_2O_3 with particle sizes less than 100 nm, one usually deals with low-temperature modifications, since the particle size for α-Al_2O_3 most often exceeds 100 nm. Laser methods are capable of providing the synthesis of Al_2O_3 with the required particle size.

In this study, the CO_2 laser vaporization technique was used for the preparation of $Al_2O_3:Eu^{3+}$ nanophosphors. The use of the laser method with its wide range of possibilities for the production of nanoparticles with different chemical compositions makes it possible to vary the properties of the obtained nanomaterials in a wide range. The most important among these possibilities is the opportunity of synthesizing nanoparticles with a size range from 1-2 nm to 100 nm. Condensation of nanoparticles in this method occurs in a flow of buffer gases at different pressures, which provides particles with different properties. To the best of our knowledge, the photoluminescent properties of laser synthesized Eu doped Al_2O_3 nanophosphors especially with the particle size below 10 nm were comprehensively studied only in our early work [6]. This work focuses on the study of quantum efficiency and color coordinates of laser synthesized $Al_2O_3:Eu^{3+}$ nanophosphors depending on synthesis conditions.

2. Experimental techniques

2.1. Synthesis of europium doped $Al_2O_3:Eu^{3+}$ nanophosphors

$Al_2O_3:Eu^{3+}$ nanophosphors were obtained by laser vaporization of α-$Al_2O_3:Eu^{3+}$ (Eu concentration is 1.0 ± 0.2 wt%, $T_{cal.} = 1250$ °C/4h) ceramic targets irradiated by a cw CO_2 laser (the power density up to ~ 0.06 MW/cm^2) with subsequent condensation of vapor in buffer gas flow in a vaporization chamber [6-8, 13-15]. To investigate the influence of synthesis conditions on PL properties of Eu^{3+}, the $Al_2O_3:Eu^{3+}$ nanophosphors were synthesized in different buffer gas: a) helium (99.99%), b) argon (99.99%), and c) helium supplemented with oxygen (99.7%). To obtain nanophosphors with the different mean sizes of nanoparticles, the gas pressure in the vaporization chamber was varied from 0.1 to 0.9 bar.

2.2. Characterization

Elemental analysis of the $Al_2O_3:Eu^{3+}$ nanophosphors was made using X-ray fluorescence spectroscopy (XRF) on an ARL – Advant'x analyzer with the Rh anode of the X-ray tube.

The morphology of the synthesized samples was characterized by transmission electron microscopy (TEM) on a JEM-2010 electron microscope.

Phase composition of the $Al_2O_3:Eu^{3+}$ nanophosphors was revealed by X-ray diffraction (XRD) analysis. X-ray diffraction patterns were recorded on a Bruker D8 diffractometer. Measurements were performed in the 2θ range of 20 – 70° with a step of 0.05° and acquisition time of 3 s. The ICDD PDF-2 database (Powder Diffraction File database PDF-2, International Center for Diffraction Data, USA, 2009) was used for phase identification.

Photoluminescence (PL) spectra were measured on a Cary Eclipse (Varian) fluorescence spectrophotometer. To measure an absolute quantum yield (QY) of the $Al_2O_3:Eu^{3+}$ nanophosphors, a Spectralon covered G8 integration sphere (GMP SA, Switzerland) was coupled to the Fluorolog 3 (Horiba Jobin Yvon) spectrofluorometer. At least three measurements of QY were made and the average value is reported. According to estimations, a consistent experimental error for the QY measurements did not exceed 15%. The color coordinates of luminescence chromaticity of the samples (in the Commission International de l'Eclairage (CIE) 1931 diagram) were calculated for the PL spectral range 565 – 725 nm. A color calculator was used to calculate the x, y color chromaticity coordinates.

3. Results and discussion

The XRF analysis reveals that after vaporization of the α-$Al_2O_3:Eu^{3+}$ targets and subsequent condensation of vapor, europium enters the composition of the produced Al_2O_3 nanoparticles with the concentration 0.92 ± 0.04 wt%. The phase composition of the synthesized $Al_2O_3:Eu^{3+}$ nanophosphors is represented mostly by γ-Al_2O_3 with the cubic structure ($Fd\bar{3}m$).

Vaporization in He and Ar buffer gas at 0.1 – 0.9 bar pressures leads to the formation of $Al_2O_3:Eu^{3+}$ nanometer particles, **Fig.1**. The sample obtained at 0.1 bar (He) contains predominantly faceted nanoparticles with the mean size of 7.2 ± 4.1 nm. An increase in the He pressure to 0.9 bar leads to growth in the average particle size to 20.9 ± 17.7 nm, while the shape of the nanoparticles becomes close to spherical. Similar changes occur when He is replaced by Ar as a buffer gas. The particle sizes obtained at a pressure of 0.1 bar (Ar) and 0.3 bar (Ar) are 14.9 ± 9.0 nm and 16 ± 9.4 nm, respectively.

PL properties of $Al_2O_3:Eu^{3+}$ nanophosphors under indirect (to charge transfer $O^{2-} \rightarrow Eu^{3+}$ transition, $\lambda_{ex} = 232$ nm) excitation are predominantly determined by Eu^{3+} ions red emission with inhomogeneously broadened bands in the region of 565-725 nm, **Fig.2**. These bands correspond to

$^5D_0 \rightarrow {}^7F_J$ (J = 0 - 4) transitions of Eu^{3+}. The hypersensitive electric dipole transition $^5D_0 \rightarrow {}^7F_2$ dominates in the spectrum and is responsible for the red emission of nanophosphors. It was shown that Eu^{3+} ions incorporate into the nanosized Al_2O_3 matrix in positions with C_{3v} local symmetry. Similar structural positions of Eu^{3+} were observed by the authors for phosphors based on α-Al_2O_3 and γ-Al_2O_3 obtained by other synthesis methods [11, 12].

The PL decay of Eu^{3+} in the region of the most intense transition $^5D_0 \rightarrow {}^7F_2$ falls within the millisecond range and is adequately described by the bi-exponential decay with "slow" and "fast" components.

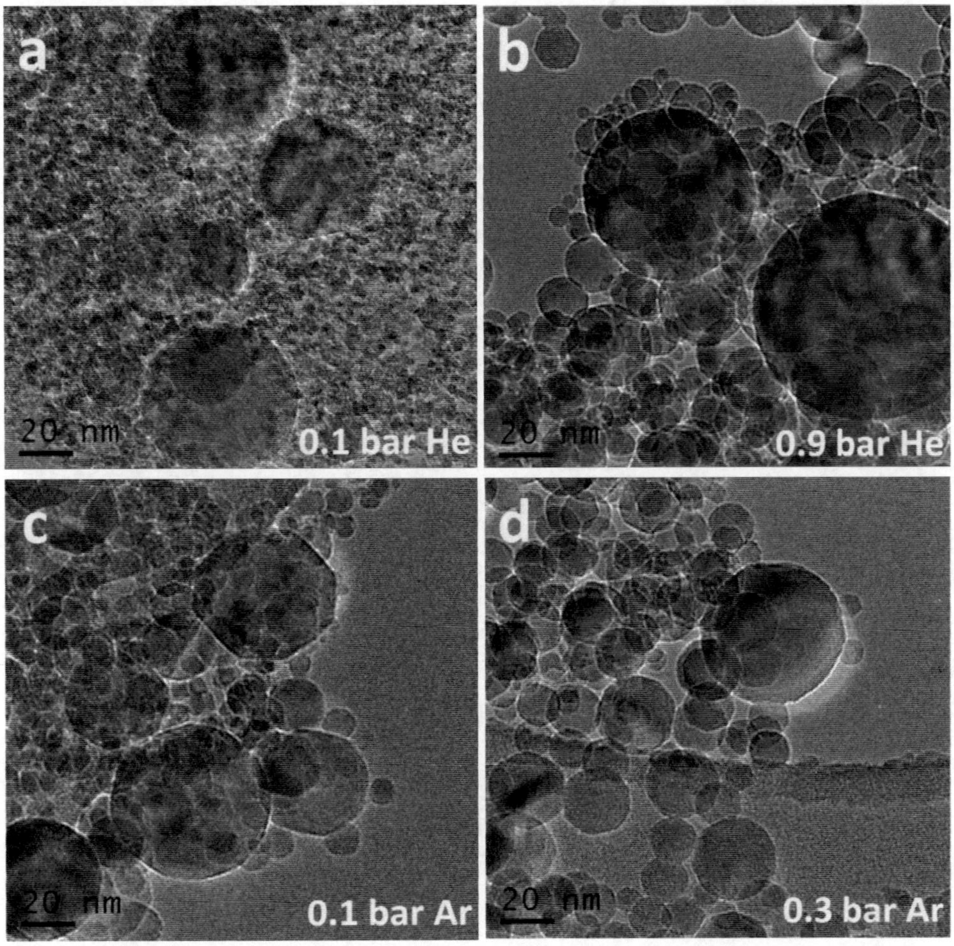

Figure 1. TEM image of the size series of Al_2O_3:Eu^{3+} nanophosphors obtained at different pressures and compositions of the ambient gas: a – 0.1 bar He, b – 0.9 bar He, c – 0.1 bar Ar, d – 0.3 bar Ar.

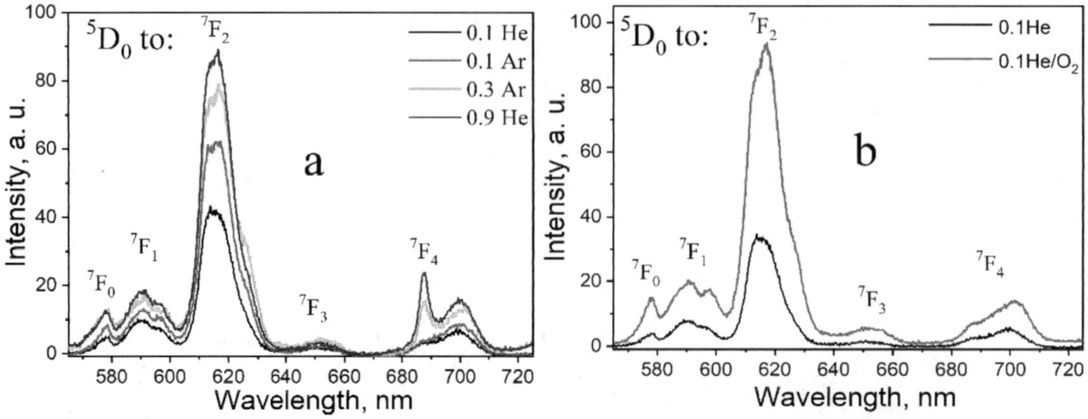

Figure 2. PL spectra of the size series of $Al_2O_3:Eu^{3+}$ nanophosphors obtained at different pressures and compositions of the ambient gas (***a***) and $Al_2O_3:Eu^{3+}$ nanophosphor obtained at 0.1 He supplemented with O_2 (***b***). All PL spectra were recorded at $\lambda_{ex} = 232$ nm.

Normalized intensities of $^5D_0 \rightarrow {^7F_2}$ transition, the calculated values of asymmetry parameter R, and lifetimes ("slow" component) are presented in **Fig. 3** as a function of the particle size.

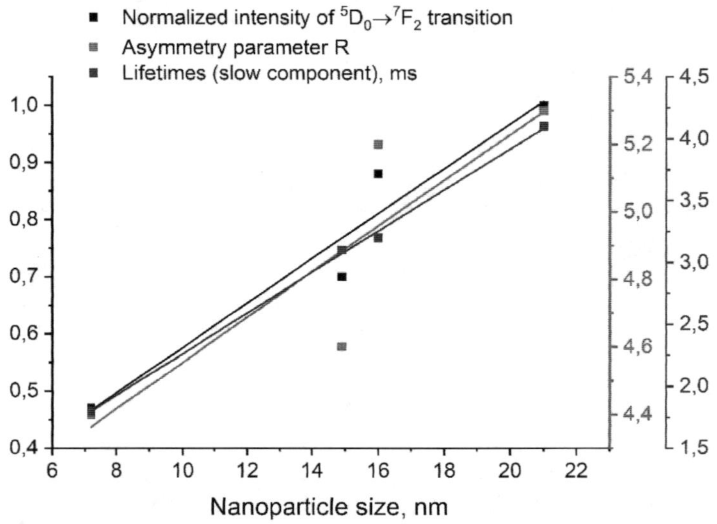

Figure 3. Normalized intensities of $^5D_0 \rightarrow {^7F_2}$ transition, asymmetry parameter R, and lifetimes ("slow" component) as a function of the particle size of $Al_2O_3:Eu^{3+}$ nanophosphors.

With a decrease in the nanoparticle size, a shortening of the PL lifetime ($\lambda_{em} = 618$ nm, $\lambda_{ex} = 232$ nm) was observed, which is due to an increase in the probability of nonradiative energy transfer from Eu^{3+} ions to adsorbed OH groups. It can be noted that the parameter R increases with increasing nanoparticle size. Thus, it can be concluded that the local symmetry of Eu^{3+} decreases with increasing nanoparticle size. In the work for $Eu^{3+}:KGd(WO_4)_2$ nanoparticles, the opposite effect of decreasing the R-value was observed [16].The authors associated this effect with the contribution of the surface. In our case, the above effects are due to the conditions for the synthesis of this material, namely the pressure and the composition of the atmosphere in the vaporization chamber. Comparison of the luminescence characteristics of $Al_2O_3:Eu^{3+}$ nanoparticles obtained in different atmospheres showed that the introduction of oxygen into the composition of the buffer gas mixture during the synthesis of the samples increases the intensity and lifetime of photoluminescence. The role of oxygen addition in the preparation of nanoparticles by laser vaporization was discussed by us earlier and is associated with a decrease in oxygen vacancies in the samples [6].

The CIE color chromaticity coordinates of nanosized $Al_2O_3:Eu^{3+}$ obtained from the PL spectra are marked on the CIE 1931 chromaticity diagram, **Fig. 4**. The values of the color chromaticity coordinates (x, y) were found to be (0.613, 0.386), (0.679, 0.320), (0.615, 0.385), (0.609, 0.391), (0.606, 0.393), and (0.704, 0.296) for $Al_2O_3:Eu^{3+}$ nanophosphors obtained at 0.1 bar (He), 0.1 bar (He+O_2), 0.1 bar (Ar), 0.3 bar (Ar), 0.9 bar (He), and 0.9 bar (He+O_2), respectively. The obtained chromaticity coordinates are close to the ideal red light (0.67, 0.33), which makes it possible to use $Al_2O_3:Eu^{3+}$ as luminescent devices and applications.

.

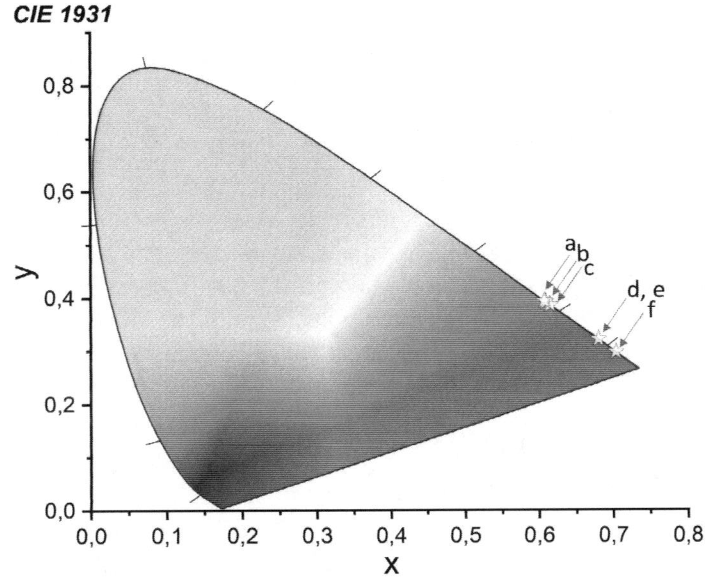

Figure 4. The CIE 1931 chromaticity diagram for $Al_2O_3:Eu^{3+}$ nanophosphors: ***a*** - 0.1 bar (He), ***b*** - 0.1 bar (He+O2), ***c*** - 0.1 bar (Ar), ***d*** - 0.3 bar (Ar), ***e*** - 0.9 bar (He), ***f*** - 0.9 bar (He+O_2).

The absolute quantum yield (QY) measured using integrating sphere pumped at 395 nm (to $^7F_0 \rightarrow {}^5L_6$ transition) for the $Al_2O_3:Eu^{3+}$ nanophosphors obtained at 0.1 bar (He), 0.1 bar (He+O_2), 0.1 bar (Ar), 0.3 bar (Ar), 0.9 bar (He), and 0.9 bar (He+O_2) was 2.8, 6.4, 3.6, 4.6, 11, and 14.3%, respectively. It can be seen that the quantum yield is maximum for particles with the largest sizes in the series. From an analysis of the luminescence decay, it was concluded that, for particles with the smallest sizes, the probability of nonradiative relaxation is maximal, which in turn leads to a decrease in the luminescence intensity and quantum efficiency. Earlier studies of the quantum efficiency of $Gd_2O_3:Eu^{3+}$ and $Y_2O_3:Eu^{3+}$ laser synthesized nanopowders showed that for particles with close sizes to $Al_2O_3:Eu^{3+}$ nanophosphors, the QY values are 4% and 21%, respectively [7, 8]. The values obtained in this work are significantly higher than for $Gd_2O_3:Eu^{3+}$ and lower than for the $Y_2O_3:Eu^{3+}$ nanophosphor. Since the QY of nanophosphors strongly depends on many factors (for example, the concentration of europium, the size of nanoparticles and their crystallinity, etc.), there is every reason to believe that, with the help of further optimization of a number of parameters, it is possible to achieve a higher QY for red $Al_2O_3:Eu^{3+}$ emission.

4. Conclusion

Laser vaporization by a cw CO_2 laser in flowing mixture of Ar/He and O_2 was used to obtain $Al_2O_3:Eu^{3+}$ nanophosphors. The particle size was varied from 7.2 ± 4.1 nm to 20.9 ± 17.7 nm by changing the gas composition and pressure in the vaporization chamber. As follows from TEM images, such particles are represented by faceted and spherically nanocrystallites. XRD studies demonstrated that the phase composition of $Al_2O_3:Eu^{3+}$ nanophosphors is represented predominantly by γ-Al_2O_3 with the cubic structure. Photoluminescence of $Al_2O_3:Eu^{3+}$ nanoparticles was studied under direct (to $^7F_0 \rightarrow {}^5L_6$ transition) and indirect (to charge transfer $O^{2-} \rightarrow Eu^{3+}$ transition) excitation with $\lambda_{ex} = 232$ and 395 nm. Luminescence properties of $Al_2O_3:Eu^{3+}$ nanophosphors are predominantly determined by Eu^{3+} ions red emission with inhomogeneously broadened bands in the region of 550-750 nm corresponding to $^5D_0 \rightarrow {}^7F_J$ (J = 0 - 4) transitions of Eu^{3+}. Hypersensitive transition $^5D_0 \rightarrow {}^7F_2$ dominates in the spectrum and is responsible for the red emission. The effect of crystallite size on luminescence intensity, decay times, color coordinates and quantum efficiency of $Al_2O_3:Eu^{3+}$ nanocrystals was observed. It was shown that the introduction of oxygen during the synthesis improves the luminescence performance. The obtained chromaticity coordinates and absolute QY (~ 14%) indicate the possibility of using red nanophosphors based on $Al_2O_3:Eu^{3+}$.

Acknowledgment

The research was supported by the Ministry of Science and Higher Education of the Russian Federation, № **121031700313-8** (M. Rakhmanova) and was financially supported by the Russian Foundation for Basic Research, Project No. **19–32–60027** (A. Kostyukov).

CRediT authorship contribution statement

A. I. Kostyukov: Conceptualization, Investigation, Writing-original draft. *A. A. Nashivochnikov*: Investigation. *M. I. Rakhmanova*: Investigation. *V. N. Snytnikov:* Conceptualization.

Declaration of competing interest

The authors declare that they have no known competing financial interests or personal relationships that could have appeared to influence the work reported in this paper.

References

[1] Fang M-H, Mahlik S, A Lazarowska, M Grinberg, M S Molokeev, H -S Sheu, J – F Lee, R -S Liu 2019 *Angew. Chem. Int. Ed.* **58** (23) 7767.
[2] Lin C C, Tsai Y-T, Johnston H E, Fang M-H, Yu F, Zhou W, Whitfield P, Li Y, Wang J, Liu R-S, Attfield J P 2017 *J. Am. Chem. Soc.* **139** 11766.
[3] Zhao M, Zhou Y, Molokeev M S, Zhang Q, Liu Q, Xia Z 2019 *Adv. Opt. Mater.* **7** 180163 1.
[4] Zhao M, Liao H, Molokeev M S, Zhou Y, Zhang Q, Liu Q, Xia Z 2019 *Light Sci. Appl.* **8** (38) 1.
[5] Liao H, Zhao M, Molokeev M S, Liu Q, Xia Z 2018 *Angew. Chem.* **57** (36) 11728.
[6] Kostyukov A I, Snytnikov V N, Zhuzhgov A V, Cherepanova S V, Ishchenko A V, Baronskiy M G, Snytnikov V N 2020 *Journal of Alloys and Compounds* **815** 152476:1-12.
[7] Kostyukov A I, Snytnikov V N, Snytnikov V N, Rakhmanova M I, Kostyukova N Y, Ishchenko A V, Cherepanova S V, Krylov A S, Aleksandrovsky A S 2021 *Journal of Luminescence* **235** 118050:1-11.
[8] Kostyukov A I, Snytnikov V N., Snytnikov V N, Ishchenko A V, Rakhmanova M I, Molokeev M S, Krylov A S, Aleksandrovsky A S 2020 *Optical Materials* **104** 109843:1-5.
[9] Baronskiy M G, Kostyukov A I, Larina T V, Snytnikov V N, Zaitseva N A, Zhuzhgov A V 2019 *Materials Chemistry and Physics* **234** 403.
[10] Rastorguev A A, Baronskiy M G, Zaitseva N A, Isupova L A, Kostyukov A I, Larina T V,

Pakhomov N A, Snytnikov V N 2014 *Inorganic Materials: Applied Research* **5** (5) 476.

[11] Kumar S, Prakash R, Kumar V, Bhalerao G M, Choudhary R J, Phase D M 2015 *Advanced Powder Technology* **26** 1263.

[12] Monteiro M A F, Brito H F, Felinto M C F C M, Brito G E S, Teotonio E E S, Vichi F M, Stefani R 2008 *Microporous and Mesoporous Materials* **108** 237.

[13] Kostyukov A I, Snytnikov V N, Rakhmanova M I, Kostyukova N Y, Snytnikov V N 2020 *Optical Materials* **110** 110508:1-6.

[14] Kostyukov A I, Baronskiy M G, Larina T V, Snytnikov V N, Zaitseva N A, Pochtar A A, Ishchenko A V, Cherepanova S V, Snytnikov V N 2020 *Materials Characterization* **169** 110664.

[15] Kostyukov A I, Snytnikov V N, Yelisseyev A P, Zhuzhgov A V, Kostyukova N Y, Ishchenko A V, Cherepanova S V, Snytnikov V N 2021 *Advanced Powder Technology* **32** (8) 2733.

[16] Pązik R, Zych A, Stręk W *2009 Materials Chemistry and Physics* **115** (1-2) 536.

Journal of Physics: Conference Series **2067** (2021) 012009 doi:10.1088/1742-6596/2067/1/012009

High-Efficiency Mini-Slab Laser Based on Tm-doped Double Tungstate Crystal

S M Vatnik[1], I A Vedin[1], M D Kolker[1], A A Pavlyuk[2]

[1] Institute of Laser Physics SB RAS, 13/3 Lavrentjeva Ave., Novosibirsk, Russia
[2] Institute of Inorganic Chemistry SB RAS, 3 Lavrentjeva Ave., Novosibirsk, Russia

E-mail: vatnik@laser.nsc.ru, vedin@laser.nsc.ru

Abstract. We report on highly-efficient room-temperature lasing in 5at.%Tm:$KLu(WO_4)_2$ mini-slabs side-pumped by a 35W diode bar. QCW (duty cycle ~ 14%) output power of 1.47 W at 1908 nm has been demonstrated with optical and slope efficiencies being of 33 and 43% respectively. In our experiments, we used samples of active elements produced in the slabs form with Brewster's angle cut faces and original laser cavity design.

1. Introduction

Lasers emitting in the spectral range of 2 μm are of particular interest for many scientific and technical applications, including remote sensing, medicine, and optical pumped mid-infrared OPO [1-4]. Crystals of double rare-earth potassium tungstate's doped with trivalent thulium ions are promising laser materials for compact highly efficient sources of coherent two-micron (1.8–2.0 μm) radiation and well known for excellent laser properties. They have rather high gain cross-sections and would allow an efficient diode pumping by widely available laser diodes and bars emitting at 800–810 nm [5-7]. It should be noted that they have about low thermal conductivity (3 W/m·K) and microhardness (~5), but in our previous works it was demonstrated the possibility of increasing the output power of thulium lasers by using active elements of a special shapes including thin disks, epitaxial and composite structures [8,9]

In previous works, we have experimentally achieved efficient multi-watt lasing in N_m-cut and *AT*-cut slab-shaped laser crystals [10]. For example, a 16.7 W maximum CW output at 1910 nm was achieved for N_m-cut slab-shaped laser crystal with a slope efficiency of 41 % and an optical-to-optical conversion efficiency of 31 %. Key past results are summarized in Table 1.

Table 1. Slope efficiency, optical efficiency and maximum output power of the different types slab active elements in the CW mode for T=10 %. AT is the "athermal" direction.

Type of mini-slab active element		Slope Efficiency, %	Optical Efficiency, %	Output power, W
AT-cut	[12]	40	23	~9.7
N_m-cut	[12]	41	31	~16.7
N_m-cut	**this work**	**43**	**33**	**~10.0**

In this paper, we present the first results of investigation of the oscillation performance of laser based on the 5at.%Tm:KLu(WO_4)$_2$ mini-slabs. In our experiments, we used samples of active elements produced in the slabs form with Brewster's angle cut faces and original laser cavity design. QCW (duty cycle ~14 %) output power of 1.47 W at 1908 nm has been demonstrated with optical and slope efficiencies being of 33 and 43 % respectively.

2. Experiment and results

The potassium lutetium double tungstate crystal, 5at.%Tm:KLu(WO_4)$_2$ was grown in the Institute of Inorganic Chemistry SB RAS, by the modified Czochralski technique (from the flux) using potassium ditungstate, $K_2W_2O_7$, as a solvent, under low temperature gradients, ~ 1 K/cm. The seed was oriented along the ***b*** crystallographic axis. A platinum crucible with a diameter of 70 mm and a length of 120 mm was used, and the pulling rate was as low as 0.1-0.2 mm/h [11].

After that, the piece of 5at.%Tm:KLu(WO_4)$_2$ crystal boule was processed into a mini-slabs with Brewster's angle cut faces having 6.2 mm in length along N_m axis by 0.7 mm along N_p (*b*-) axis and by 0.36 mm along N_g axis, all facets are flat and polished, see Fig. 1a. The bottom (upper) 0.36×6.2 mm^2 facet was HR (AR) coated for pump wavelengths centred at 808 nm, Brewster's angle cut end-faces were not AR coated for lasing wavelengths. At the final stage the coated slab was clamped between two copper heatsinks with an indium solder, see inset to figure 2a.

The active element was pumped by the collimated radiation of commercial CS-type diode bar with the total optical power up to 35 W at a wavelength of 808 nm. The total optical loss related to delivery of pump light to the upper slab facet was estimated to be ~5 %, the rectangular pump spot was of 6.0 mm in length by 0.1 mm in width allowing efficient side pumping. In our case, a ten-pass diode pumping scheme was used, quite similar that of [10]. The residual pump power was not measured, because we established that all pumping was absorbed in the active element in ten passes, see [10] for more details.

The temperature of the copper base, on which all the elements of the optical circuit and the radiator of the active element were attached, was maintained constant (T = 25±0.5 °C) using modules of the thermal stabilization system based on Peltier elements.

The laser cavity was formed by a spherical rear mirror with a radius of curvature equal to -40 mm, AR-coated YAG crystal lens with a focal length of 20.0 mm and flat output coupler. The distance between the rear mirror and output coupler was approximately 80 mm, see Fig. 1b. The transmission output coupler was 10 % in the spectral range 1850–1950 nm.

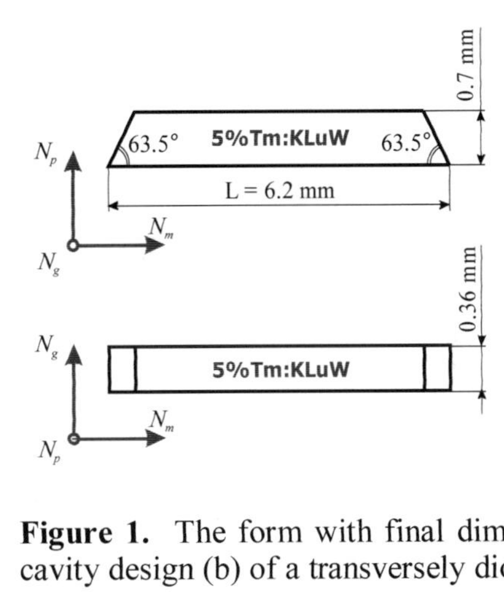

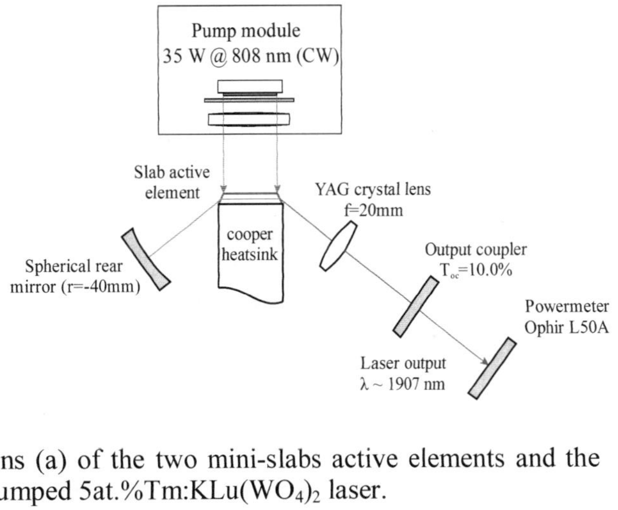

Figure 1. The form with final dimensions (a) of the two mini-slabs active elements and the cavity design (b) of a transversely diode pumped 5at.%Tm:KLu(WO_4)$_2$ laser.

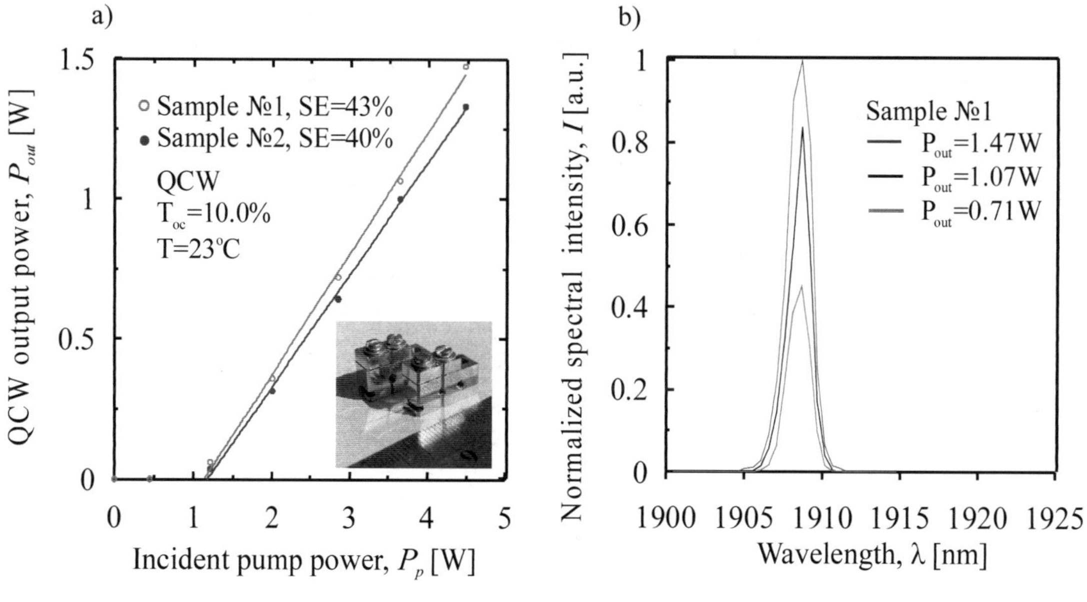

Figure 2. Input-output dependences of N_m-cut 5at.%Tm:KLu(WO_4)$_2$ slabs in QCW mode (a) and typical normalized output spectrum for different output power (b). The inset shows the active elements. SE is the slope efficiency.

Figure 1a shows the dependences of the output laser power on the QCW diode pump power (pump pulse duration 7 ms, pulse repetition rate 20 Hz, duty cycle 14%) for two identical samples of mini-slabs. The optical powers of the light beams (pumping and lasing) were measured with powermeter Ophir L50A. All output spectrum emission patterns (see figure 1b) were recorded by MDR-204 monochromator equipped with a photoresistor FR-185 and a Unipan-233 selective nanovoltmeter as a preamplifier. The spectral resolution was ~ 0.5 nm. Similarly to [10], the emission of the N_m-cut slabs with Brewster's angle cut end-faces was centered at 1908 nm with a spectral bandwidth (full width at half maximum, FWHM, of ~ 2 nm).The average output power over 1.4 W at 1908 nm has been demonstrated for all samples with slope efficiencies being of over 40 % with respect to the incident pump power and an optical-to-optical conversion efficiency of 33 %. At this stage, CW mode testing has not been carried out but these results are consistent with calculated output power about ~10 W on the CW mode accounting duty cycle 0.14 (7 ms to 50 ms).

3. Summary

In summary, a high-efficiency QCW laser performance of Brewser-cut N_m 5at.%Tm:KLu(WO_4)$_2$ slab seems to be very similar to other Tm^{3+}-doped double tungstate gain elements including thin disks and epitaxial structures that have been studied previously. The further work will be focused on the increasing laser output power up to 10…20 W in CW mode by combining two active elements into a single resonator. The further optimization of laser cavity parameters should improve laser beam quality up to $M^2<1.5$, and active Q-switching regime with KTP/RTP electooptics is also under intensive study.

References

[1] Fischer C, Sorokin E, Sorokina I, Sigrist M 2005 Photoacoustic monitoring of gases using a novel laser source tunable around 2.5μm *Opt. Lasers Eng.* **43** 573

[2] Henriksson M, Tiihonen M, Pasiskevicius V, Laurell F 2006 $ZnGeP_2$ parametric oscillator pumped by a linewidth-narrowed parametric 2 μm source *Opt. Lett.* **31** 1878

[3] Scholle K, Lamrini S, Koopmann P, Fuhrberg P 2010 2 μm Laser Sources and Their Possible Applications, *Source: Frontiers in Guided Wave Optics and Optoelectronics* 21 471
[4] John D, Craig S 1992 2-μm laser applications *Proc. SPIE* **1627** 1
[5] Demidovich A, Kuzmin A, Nikeenko N, Titov A, Mond M, Kueck S 2002 Optical characterization of Yb,Tm:KYW crystal concerning laser application *Journal of Alloys and Compounds* **341** 124
[6] Mateos X, Petrov V, Liu J, Pujol M, Griebner U, Aguilo M, Diaz F, Galan M, Viera G 2006 Efficient 2-μm continuous-wave laser oscillation of Tm^{3+}:$KLu(WO_4)_2$ *IEEE J. Quantum Electron.* **42** 1008
[7] Petrov V, Pujol M, Mateos X, Silvestre O, Rivier S, Aguilo M, Sole R, Liu J, Griebner U, Diaz F 2007 Growth and properties of $KLu(WO_4)_2$, and novel ytterbium and thulium lasers based on this monoclinic crystalline host *Laser & Photon. Rev.* **1** 179
[8] Vatnik S, Vedin I, Kurbatov P, Pavlyuk A 2014 Efficient 2-mm laser oscillation of 5%Tm^{3+}:$KLu(WO_4)_2$ disks and 5%Tm^{3+}:$KLu(WO_4)_2/KLu(WO_4)_2$ composite structures *Quantum Electronics* **44** 989
[9] Vatnik S, Vedin I, Segura M, Mateos X, Pujol M, Carvajal J, Aguiló M, Díaz F, Petrov V, Griebner U 2012 Efficient thin-disk Tm-laser operation based on Tm:$KLu(WO_4)_2/KLu(WO_4)_2$ epitaxies *Opt. Lett.* **37** 356
[10] Bagayev S, Vatnik S, Vedin I, Orlovich V, Kuleshov N, Pavlyuk A, Gusakova N, Kurilchik S, Yasukevich A, Kisel V, Yumashev K, Loiko P, Dashkevich V 2019 Efficient Tm-laser operation based on 5at.%Tm:$KLu(WO_4)_2$ with N_m and at orientations *AIP Conference Proc.*, **2098**, 020003
[11] Vatnik S, Pujol M, Carvajal J, Mateos X, Aguiló M, Díaz F, Petrov V 2009 Thermo-optic coefficients of monoclinic $KLu(WO_4)_2$ *Appl. Phys.* **B95**, 653
[12] Pavlyuk A, Vasiliev Y, Kharchenko L, Kuznetsov F 1993 Low thermal gradient technique and method for large oxide crystals growth from melt and flux *Proc. of APSAM-92* **1** 164

Journal of Physics: Conference Series 2067 (2021) 012010 doi:10.1088/1742-6596/2067/1/012010

Fluorescence quantum efficiency of $Ti^{3+}:Al_2O_3$ at cryogenic temperatures and excitation by laser diodes

A Gribanov, G Nikolaev, M Mosin, D Yakovin, M Yakovin

Institute of Automation and Electrometry of SB RAS, Novosibirsk, 630090 Russia

E-mail: m.d.yakovin@mail.ru

Abstract. In this work the temperature dependences of the fluorescence of a titanium-sapphire crystal excited by radiation with λ=507 nm, λ=454 nm and λ=405 nm are measured. The quantum efficiency of fluorescence is determined. It was found for the first time that the quantum efficiency of fluorescence exceeds 100% in the cryogenic temperature range when the crystal is excited by radiation with λ = 454 nm. It is equal to (117 ± 5) % and (114 ± 5) % at T = 77 K for crystals grown by the Czochralski method and by gradient of temperature one, respectively.

1. Introduction

It is known that at cryogenic temperatures the thermal conductivity of sapphire sharply increases and its spectroscopic characteristics change, which makes it possible to use high pump intensities and provide the highest output power of generation of titanium-sapphire lasers and aplifiers.

Measurements of the quantum efficiency (QE) of the titanium-sapphire (TiSa) crystal were carried out by various methods. For example, in [1], the internal QE was measured in the generation of a titan-sapphire laser at room temperature (64%). When a TiSa laser was pumped at liquid nitrogen temperature by the blue lines of an argon-ion laser, the quantum efficiency of such laser was ~100% [2]. In [3], qualitative indications are given that at cryogenic temperatures and when a TiSa crystal is pumped at wavelengths of ~ 450 nm, the QE of fluorescence is greater than 100%.The authors of [4] performed measurements of the fluorescence quantum efficiency in a wide temperature range (150–550 K), when excited by a laser with a wavelength of 543 nm. As far as we know, at cryogenic temperatures there are no quantitative measurements of the quantum efficiency of fluorescence of a titanium-sapphire crystal irradiated with light radiation with a wavelength in the range of 400 - 500 nm.

The issue of quantum efficiency at cryogenic temperatures is very topical and is of scientific interest from the point of view of the processes occurring during excitation of a TiSa laser, as well as when choosing the optimal pump wavelength for TiSa lasers and amplifiers at low temperatures.

The aim of our work is to quantitatively measure the fluorescence quantum efficiency of $Ti^{3+}:Al_2O_3$ crystal at cryogenic temperatures and excitation by laser diodes at wavelengths 507 nm, 454 nm and 405 nm.

2. Experiment setup and methods

The experimental setup is shown in Figure 1. Linearly polarized radiation from a laser diodes (LD) with wavelengths of 507 nm, 454 nm and 405 nm is collimated by telescope (x0.3) to diameter of ≈1 mm and directed in $Ti^{3+}:Al_2O_3$ crystal. The fluorescence from the crystal is collected by the lens (L) and recorded by the photodetector (FD). Unabsorbed radiation emerging from the crystal is reflected from an interference filter (F), (transmission 600-950 nm) to a photodetector (FD) to control the absorbed power.

The unsaturated gain g_0 under these excitation conditions was 10^{-5} cm^{-1} (LD power ~30–100 mW), and the contribution of stimulated radiation was no more than 0.2 % of the fluorescence power incident on the photodiode.

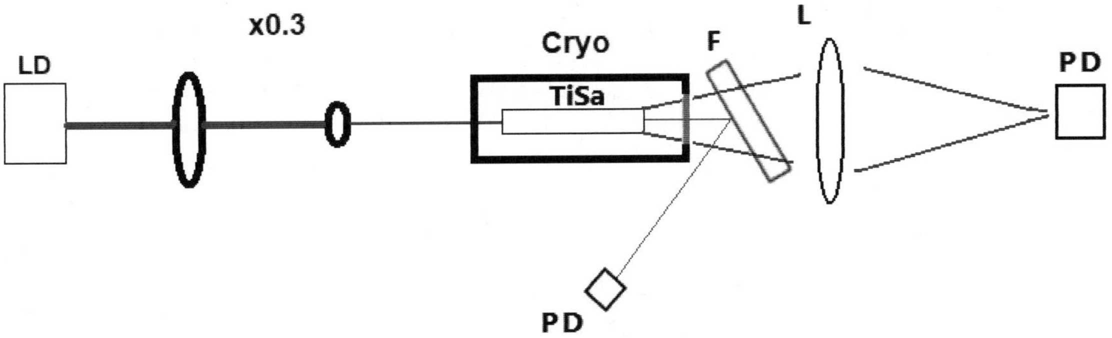

Figure 1. Experimental setup. LD – laser diode, x0.3 – collimator, Cryo – cryostat F – interference filter, L – collected lens, PD – photodetectors, TiSa – titan-sapphire crystal.

The Ti^{3+}:Al_2O_3 crystalline rod grown by the Czochralski method with a diameter of 5 mm and a length of 10 mm with Brewster-cut and optically polished facets was placed in cryostat (Cryo) whose temperature could be changed from 77 K to 320 K. Fluorescence was excited by pumping from an appropriate laser diode into the π absorption band.

Figure 2 shows the plot of crystal absorption as a function of wavelength (points) and its approximation (yellow) with Gaussian curves (red).

To determine the quantum efficiency under pumping with radiation from 507 nm to 454 nm or 405 nm, the corresponding curve of the dependence of the fluorescence intensity as a function of temperature was recorded. The temperature dependence of the fluorescence at excitation at 454 nm and 405 nm was compared with the "reference" dependence obtained at excitation at 507 nm. The data obtained at 507 nm are normalized to the value of the fluorescence quantum efficiency (69.5%) at room temperature, measured by the calorimetric method in [5]. The output powers of the lasers are selected so that the same number of photons are absorbed in the Ti^{3+}: Al_2O_3 crystal (28 mW, 38 mW and 100 mW for 507 nm, 454 nm and 405 nm respectively).

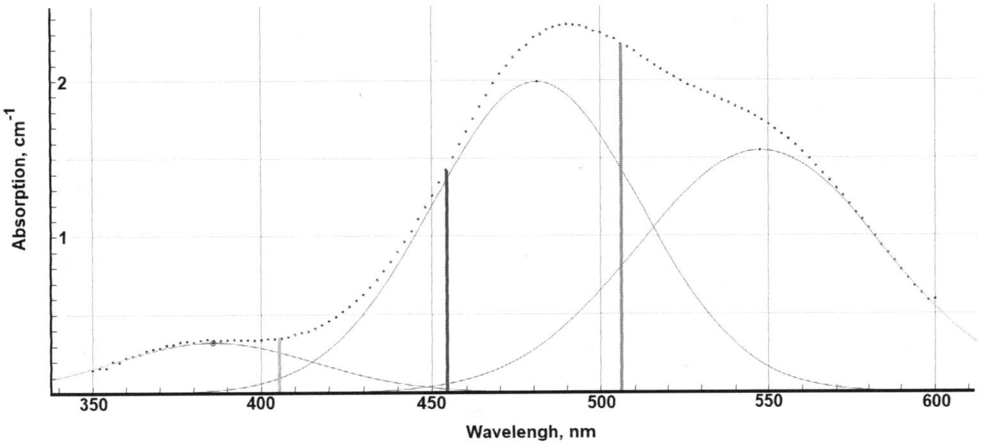

Figure 2. Optical absorption spectra (points) and corresponding fitting results of Ti:Al_2O_3 crystal with Gaussian curves. Violet, blue and green lines – wavelength of laser diodes.

3. Results and Discussion

Figure 3 shows a plot of the dependence of the fluorescence intensity of titanium ions in sapphire under excitation by radiation with λ = 507 nm, 454 nm and 405 nm as function of temperature.

The fluorescence quantum efficiency of Ti^{3+}:Al_2O_3 crystal when pumped at 507 nm was (97±5) %, and when pumped at 454 nm it was (114±5) %. The quantum efficiency under pumping at 405 nm practically did not change with temperature and amounted to about 24% ± 5%. Similar measurements were carried out for a titanium-sapphire crystal grown by the temperature gradient method (TGT). The quantum efficiency at pumping at 507 nm and 454 nm was 95% and 114%, respectively. The quantum efficiency when pumped at 405 nm was 23%.

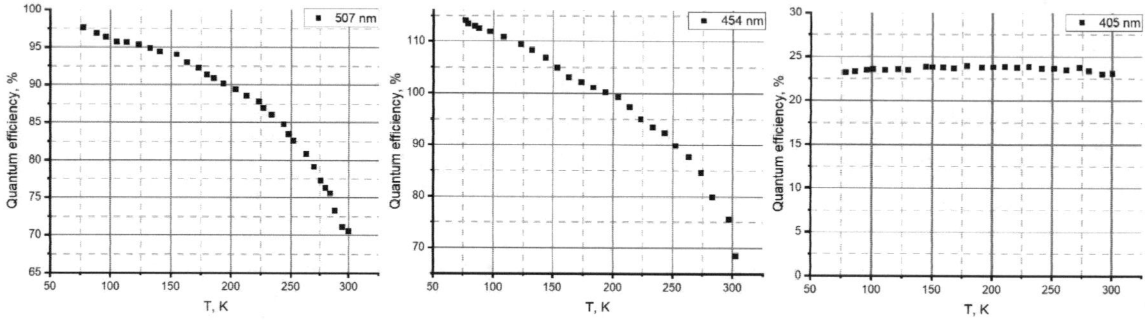

Figure 3. Quantum efficiency of the Ti^{3+}:Al_2O_3, pumped by 507 nm, 454 nm and 405 nm, as function of temperature.

The reasons for the quantum efficiency of more than 100% can be associated with the presence in the Al_2O_3 matrix of complexes of Ti^{3+} -Ti^{3+}, $2Ti^{3+}$-V_O+$2e^-$ ions, which form their own energy level schemes. It was noted in [6] that the Ti^{3+}–Ti^{3+} complex participates in the formation of absorption spectra in the 400–450 nm region. In our opinion, this can lead to an increase in the quantum efficiency of luminescence by more than 100% by transferring excitation from the energy level of this pair to the energy levels of two Ti^{3+} ions.

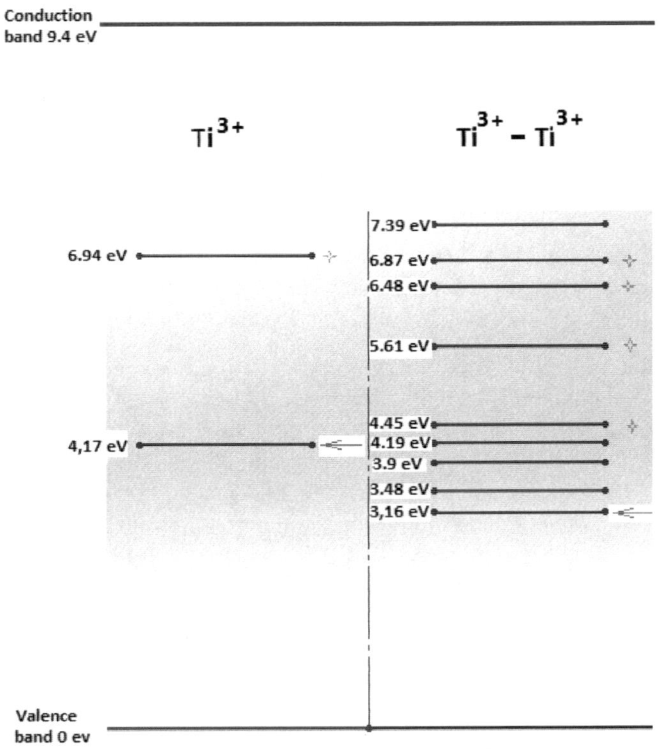

Figure 4. Simplified energy-level diagrams for isolated Ti^{3+} and complex ions of Ti in sapphire (from Ref. [7]).

The theoretically calculations of such energy levels complexes Ti^{3+}-Ti^{3+} and other complexes were theoretically performed in [7]. Figure 4 shows simplified level diagrams based on the results of this work (Figure 9b and Figure 10a). The red arrow points to the lowest inhabited level

The transition wavelengths and energies are as follows (green stars in Figure 4):

T^{3+} - 448 nm (2.77 eV),

T^{3+}-T^{3+} -334 nm (3.71 eV); 373 nm (3.32 eV); 506 nm (2.45 eV); 961 nm (1.29 eV).

It is known that theoretical calculations of energies based on approximations like [7] give a large uncertainty (up to 40%) both when calculating the energy of the lower level and when calculating the distances between them; therefore, it is difficult to accurately identify the level, perhaps this is a transition from wavelength 334 or 373 nm.

Unfortunately, there are no published data on the calculations of the energy levels of the $2Ti^{3+}$-V_O+$2e^-$ complex. In [8] the authors associate an absorption peak at about 400 nm with this complex.

4. Acknowledgments

This work was performed using the equipment of the Multiple-access centre "High-resolution spectroscopy of gases and condensed matters" in IA&E SB RAS, Novosibirsk, Russian Federation.

References

[1] Sanchez A, Fahey R, Strauss A and Aggarwal R 1986 *Opt. Lett.* **11** 363
[2] Donin V, Ivanov V, Kovalevskii V and Yakovin D 1995 *Opt. Commun.* **122** 40
[3] Backus S, Kirchner M, Lemons R, Schmidt D, Durfee C, Murnane M and Kapteyn 2017 *Opt. Express* **25** 3666
[4] Shirakov A, Burshtein Z, Shimony Y, Frumker E and Ishaaya A 2019 *Nature, Scientif. Rep.* **9**

18810
[5] Li Y, Duncan I and Morrow T 1992 *J. of Luminesc.* **52** 275
[6] Moulton P, Cederberg J, Stevens K, Foundos G, Koselja M and Preclikova J 2019 *Opt. Mat. Express* **9** 2216
[7] Kravchenko L and Fil D 2018 *J. Appl. Phys.* **123** 023104
[8] Kryvonosov Ie and Lytvynov L 2012 *Crystall. Reports* **57** 967

Millimetre-wave range optical properties of BIBO

N A Nikolaev[1,2], A A Mamrashev[2], V D Antsygin[2], D M Ezhov[3], D M Lubenko[4], V A Svetlichnyi[3], Yu M Andreev[3,4,5] and V F Losev[4]

[1]Institute of Laser Physics SB RAS, 15 B Academician Lavrentiev Ave., Novosibirsk, 630090, Russia
[2]Institute of Automation and Electrometry SB RAS, 1 Academician Koptyug Ave., Novosibirsk, 630090, Russia
[3]Tomsk State University, 36 Lenin Ave., Tomsk, 634050 Russia
4Institute of High Current Electronics SB RAS, 2/3 Akademichesky Ave., Tomsk [6]34055, Russia
[5]Institute of Monitoring of Climatic and Ecological Systems SB RAS, 10/3 Academichesky Ave., Tomsk 634055, Russia

Corresponding author: Nazar Nikolaev, email: Nazar@iae.nsk.su

Abstract. We present the thorough studies of optical properties of BiB_3O_6 (BIBO) crystal in the millimeter-wave (subterahertz) range. We observe a large birefringence $\Delta n = n_Z - n_X = 1.5$ and the values of absorption coefficients of all three axes to be less than 0.5 cm^{-1} at the frequency of 0.3 THz. The difference from visible range in angle ϕ between the dielectric axis z and crystallophysical axis X is found to be more than 6°. The simulated phase-matching curves in the xz plane of the crystal show the optimal value of the angle θ to be around 25.5°±1° for an efficient millimeter-wave generation under the pump of 1064 nm laser radiation.

1. Introduction

The negative biaxial crystal of bismuth triborate belongs to the non-centrosymmetric monoclinic space group $C2$. BIBO was demonstrated as an efficient ultraviolet radiation source, a second harmonic generator, and a parametric frequency converter of ultrashort pulses [1]. Its dielectric x-axis is parallel to the crystallophysical Y-axis (crystallographic b-axis) and the angle ϕ between the axes z and X is wavelength-dependent. Its value is about 47° in the main transparency window [2]. Previously reported that the crystal is transparent in the range below 2 THz and exhibits significant birefringence [3]. However, due to the small thickness of the samples, it was not possible to get a reliable measure of the absorption coefficients. In this work, samples with a thickness of 5 mm were studied, which made it possible to refine the absorption coefficient of the crystal and obtain some new data on the properties of the BIBO at terahertz frequencies.

2. Methods and samples

Terahertz properties were measured using a conventional terahertz time-domain spectrometer (THz-TDS). The description of the experimental setup can be found elsewhere [4]. In the current study, we applied a newly developed measurement procedure described in [5]. It allowed us to measure the absorption coefficient and the refractive index of the crystal for both axes during one cycle (without removing and rotating the sample) under nearly the same conditions. THz signals were acquired with a time step of 125 fs in the 60 ps time range which corresponded to a spectral resolution of about 20 GHz.

3. Experimental results

3.1. Optical properties

The absorption coefficient in the subterahertz range of spectra (below 0.3 THz) was found to be less than 0.5 cm^{-1} for all axes as it shown in figure 1. The absorption coefficients increase at higher frequencies showing typical behavior for the most of nonlinear crystals.

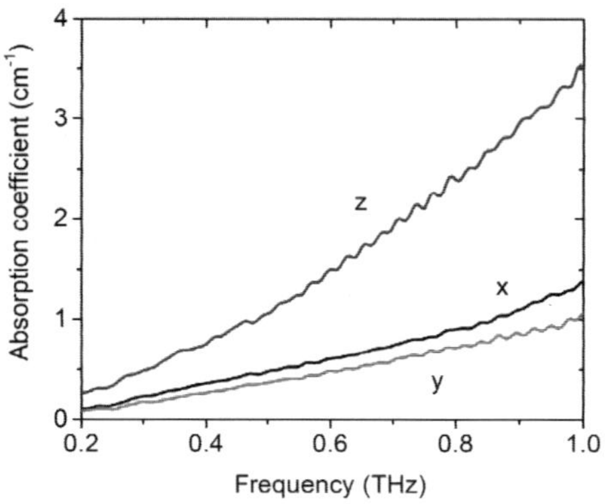

Figure 1. Absorption coefficient components of BIBO crystal in the subterahertz range.

We have measured ϕ angle using TeraScan frequency-domain spectrometer from Toptica Photonics (linewidth is about 10 MHz). The measurements were carried out with the crossed high-quality polarizers. It was shown that the dielectric frame *xyz* in the sub-THz is rotated more than 6 degrees from the visible and its dispersion was about 5 degrees (figure 2).

The dispersion of refractive indices decreases at sub-terahertz frequencies and almost disappears approaching millimeter waves (figure 2). In this area (~0.2 THz), the refractive index components are: $n_X = 2.4$, $n_Y = 2.6$, $n_Z = 3.9$. The *z*-axis shows the greatest dispersion, which is due to the influence of the same strong absorption mode.

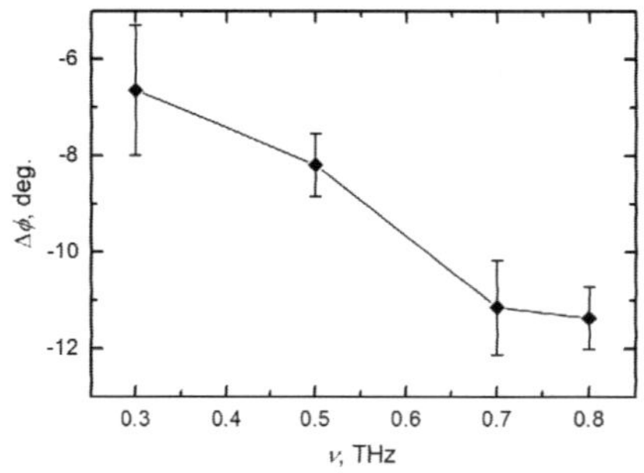

Figure 2. The difference from visible range in angle ϕ between the axes *z* and *X*.

Taking this into account, the refractive indices were refined as it shown on a figure 3. We obtain the same order of refractive indices as earlier [3], but more precisely at the frequencies below 0.2 THz due to increased SNR.

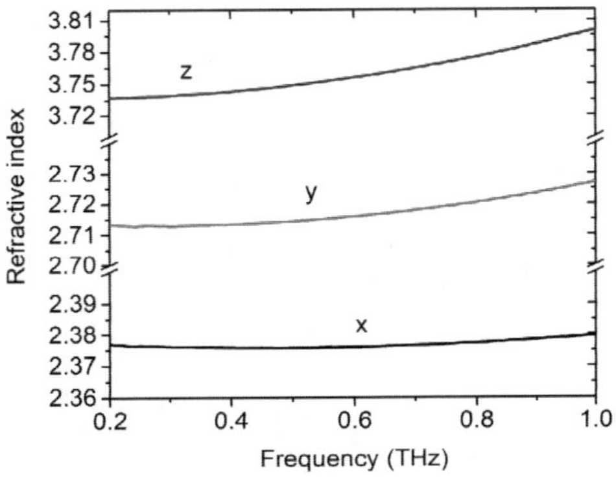

Figure 3. Refractive index of BIBO crystal in the subterahertz range.

3.2. Phase-matching for a difference frequency generation
The fulfillment of phase-matching conditions for DFG is found to be possible in the main optical plane *xz* ($\varphi = 0°$). In total, two types of three-wave interactions (out of eight possible) were discovered - *sfs* (figure 4) and *ffs* (figure 5), where *f* is fast wave, *s* is slow wave, and the first letter corresponds to the longest wavelength (the ratio of the lengths of the interacting waves is determined by the agreement $\lambda_1 \geq \lambda_2 > \lambda_3$). The figures show the dependence of the phase-matching angle θ_{pm} on the generated wavelength. The curves for the case when the second wave is fixed (1064 nm, while the third wave is shorter) do not differ from the case when the third wave is fixed (while the second wave is longer).

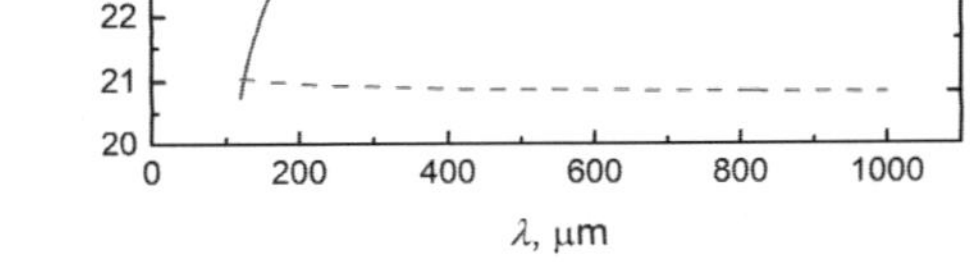

Figure 4. The dependence of the phase-matching angle on the generated wavelength via *sfs* type of interaction in the *xz* principal plane.

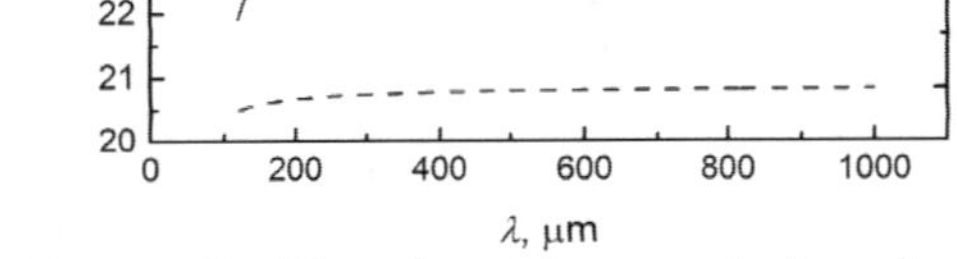

Figure 5. The dependence of the phase-matching angle on the generated wavelength via *ffs* type of interaction in the *xz* principal plane.

4. Summary
In comparison with well-known β-BBO, or LBO crystals BIBO shows the highest nonlinear coefficients and the lowest absorption in the THz range, which in turn positions it as a promising downconverter of high-power laser radiation. A found dispersion of the ϕ angle should be considered

when generating broadband terahertz waves. The optimal value of the angle θ_{pm} found to be around 25.5°±1° for the efficient millimeter-wave generation under 1 μm laser pump.

Acknowledgments

This work was supported by the Russian Science Foundation (RSF), project № 19-19-00241.

References

[1] Petrov V et al., 2010 Femtosecond nonlinear frequency conversion based on BiB3O6 *Laser & Photon. Rev.*, **4**, 53-98
[2] Hellwig H, Liebertz J, Bohaty L 2000 Linear optical properties of the monoclinic bismuth borate BiB3O6 *J. Appl. Phys.*, **88**, 240–244
[3] Li Y et al. 2020 Optical properties of BiB3O6 in the terahertz range *Results in Physics*, **16**, 102815.
[4] Rybak A et al. 2021 Terahertz Optical Properties of $KTiOPO_4$ Crystal in the Temperature Range of (− 192)–150° C, *Crystals*, **11(2)**, 125
[5] Mamrashev A et al. 2021 Terahertz Time-Domain Polarimetry for Principal Optical Axes of Anisotropic Crystals, *Photonics* **8(6),** 213
[6] Antsygin V D et al. 2013 Optical properties of borate crystals in terahertz region *Optics Communications* **309**, 333-337
[7] Nikolaev N A et al. 2018 Terahertz optical properties of LBO crystal upon cooling to liquid nitrogen temperature, *Quantum Electronics* **48(1)**, 19

Journal of Physics: Conference Series **2067** (2021) 012012 doi:10.1088/1742-6596/2067/1/012012

Detection of THz waves in GaSe:S crystals by femtosecond laser radiation with a telecom wavelength of 1.55 μm

O N Shevchenko[1,2], N A Nikolaev[1,2] and K A Kokh[1,3]

[1] Novosibirsk State University, 1 Pirogova St., 630090, Novosibirsk, Russia
[2] Institute of Automation and Electrometry SB RAS, 1 Academician Koptyug Ave., 630090, Novosibirsk, Russia
[3] Institute of Geology and Mineralogy SB RAS, 3 Academician Koptyug Ave., 630090, Novosibirsk, Russia

Corresponding author: Shevchenko Olesya; email: o.shevchenko@nsu.ru

Abstract. In this work, we study $Ga_{50\%}Se_{50-x\%}S_x$ crystals (where x = 0, 1.5, 6, 8, 11) as an electrooptic detector of terahertz pulses using probing femtosecond laser radiation with the wavelength of 1.55 μm. It was found that the sample with x = 6 provides the highest detection efficiency. The efficient value of the electrooptic coefficient of GaSe:S crystals is estimated to be about twice higher than those of GaAs in the same conditions.

1. Introduction

The progress in the technologies of optical telecommunication systems is accompanied by an increase in the modulation frequencies up to hundreds of GHz related to the so-called millimeter-wave or subterahertz range. New concepts based on the optical-to-optical conversion (nonlinear interaction) are proposed for detecting these frequencies as opposed to the classical approach of conventional electronics. Thus, the search and study of suitable nonlinear media are currently of great interest.

In recent years gallium selenide (GaSe) is attracting more attention as a material for integrated optics [1]. It is a well-known nonlinear medium and has been actively studied for the interaction of IR with millimeter- and terahertz waves including generation and detection [2-10]. A significant number of publications devoted to these tasks dedicated to the conversion of wavelengths in the vicinity of 0.8 μm [2-5], 1 μm [6-7], 2 μm [8] and 10 μm [9-10]. However, to the best of our knowledge, there are no reports devoted to the telecom wavelength of about 1.5 μm.

In this work, we present the study of GaSe crystals doped with a sulfur element (GaSe:S) as the detector of the THz waves gated by femtosecond laser pulses at a wavelength of 1.55 μm.

2. Experimental Setup

The study of the detection efficiency in GaSe:S crystals was carried out in the terahertz-time domain spectroscopy setup (THz-TDS) (figure 1). An Er-fiber laser acts as a source of femtosecond pulses at the wavelength of 1.55 μm. The pulse duration and the repetition rate are 80 fs and 77 MHz correspondingly. The output laser beam having the power of about 220 mW is divided into a pump (90%) and a probe (10%) beams using a beam-splitting cube. The pump beam passes through the acousto-optical modulator (AOM). The modulation frequency is locked with the internal generator of the lock-in amplifier. The modulated signal is transmitted to a THz generator represented by a p-InAs crystal in a magnetic field of 0.8 T. Generated terahertz pulses are collimated by an off-axis parabolic mirror and pass through a low-frequency filter that cuts off residual laser radiation. The probe beam passes through an optical delay line consisting of a motorized linear stage with a retroreflector mounted on it and enters the electrooptic crystal detector GaAs simultaneously with the THz pulse.

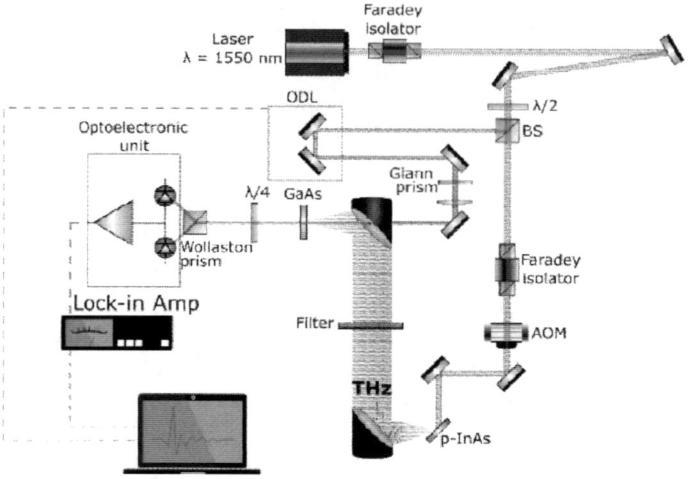

Figure 1. The scheme of the experimental setup.

The detection is based on the electro-optical Pockels effect [11]. The signal from the optoelectronic unit is acquired using the lock-in amplifier. The digitized signal is recorded by a PC

3. Samples preparation

The $Ga_{50\%}Se_{50-x\%}S_{x\%}$ samples (where x takes the values: 0, 1.5, 6, 8, 11) were grown from melts using the vertical Bridgman method using the rotation of the thermal field. The atomic concentration of sulfur was determined by the energy dispersive X-ray spectrometer combined with the scanning electron microscope. Samples were prepared by cleaving of two thicknesses: ~ 1 mm to measure the terahertz refractive index and ~ 0.3 mm to measure the detection efficiency of terahertz radiation following the estimated coherence length.

4. Selected results

4.1. IR and terahertz optical properties

Measurements of the optical refractive index of GaSe:S samples of various degrees of doping were carried out on the Metricon system at a wavelength of 1547 nm. The claimed measurement error is 0.003. The measurement results are presented in Table 1.

Table 1. The measured refractive index of $Ga_{50\%}Se_{50-x\%}S_{x\%}$ crystals at the wavelength of 1547 nm.

x	n_{1547}
0	2,75
1,5	2,744
6	2,733
8	2,713
11	2,703

The terahertz refractive index of the samples was measured on a THz-TDS with a high dynamic range. The description of the spectrometer can be found elsewhere [12-13]. The results are shown in figure 2.

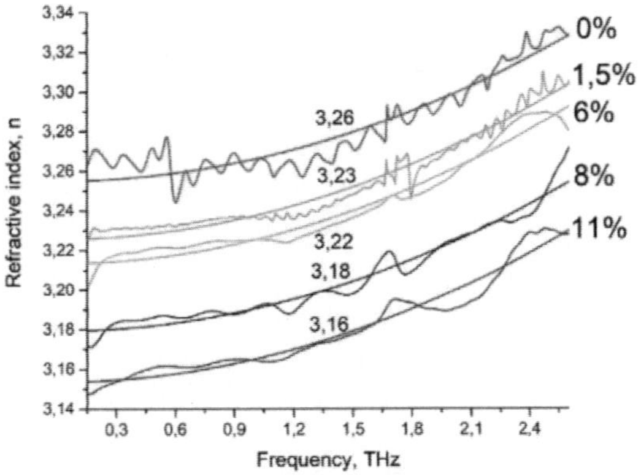

Figure 2. Terahertz refractive index of $Ga_{50\%}Se_{50-x\%}S_{x\%}$ samples for the ordinary wave, where x = 0, 1.5, 6, 8, 11.

Presented results show that the inclusion of sulfur elements in the structure of the GaSe crystal leads to the decrease of the refractive index in the IR and THz range.

4.2. Terahertz waves detection efficiency

GaAs detection crystal (figure 1) was replaced by the GaSe:S samples to study the efficiency of terahertz radiation detection. Figure 3 shows the acquired waveforms of THz pulses. It can be seen that the sample with x = 6 shows the highest signal amplitude.

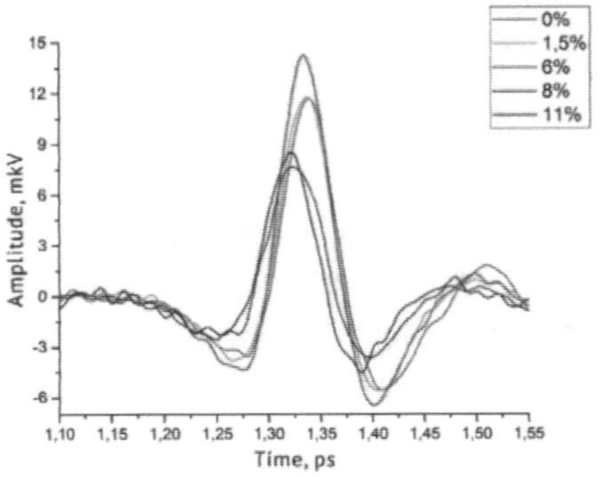

Figure 3. Detected terahertz pulses waveforms in the GaSe:S samples with different sulfur atomic concentrations.

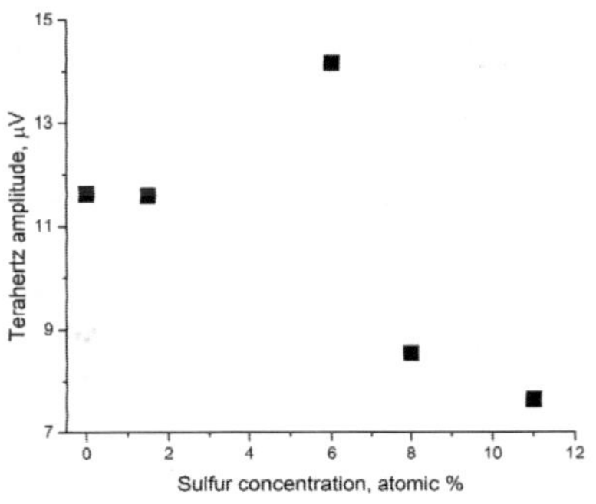

Figure 4. The peak amplitude of the THz signal dependence on the sulfur atomic concentration.

A Scatter plot of the THz peak amplitude dependence (figure 4) on sulfur atomic concentration shows that 6% is most likely not the optimal value which most probably is in the range from 3% to 6%.

4.3. Electrooptical coefficient of GaSe:S samples

Since we only change the crystal in the detection system and the terahertz wave with probing IR pulse are remains the same amplitude and energy, we can compare the electrooptic coefficient of the GaSe:S crystals with those of GaAs for which it is well-known. Considering the length of the crystals, THz, and IR refractive index we can estimate the electrooptical coefficient:

$$r_{GaSe} = \frac{L_{GaAs} n_{GaAs}^3 r_{GaAs} (1-\rho_{GaAs_{THz}})(1-2\rho_{GaSe_{opt}})}{L_{GaSe} n_{GaSe}^3 (1-\rho_{GaSe_{THz}})},$$

where L – crystal thickness, n_{GaSe}, n_{GaAs} – IR refractive indices, ρ_{opt} и ρ_{THz} – optical and terahertz reflection coefficients, respectively. In our case r_{GaAs} = 1.5 pm/V, n_{GaAs} = 3.4, and n_{GaSe} is taken from the table 1. The results are shown in Table 2.

Table 2. Electrooptical coefficient of GaSe:S crystals.

x	r_{22} [pm/V]
0	2,80
1,5	2,86
6	2,93
8	3,06
11	3,11

5. Summary

In this paper, the efficiency of terahertz waves detection in $Ga_{50\%}Se_{(50-x)\%}S_{x\%}$ crystals (where x takes the values 0, 1.5, 6, 8, and 11) is investigated using femtosecond laser pulses at the wavelength of 1.5 μm. The refractive indices of the samples in the IR and terahertz ranges were measured. It was found that the sample with x = 6 provides the highest detection efficiency. The efficient value of the electrooptic coefficient of GaSe:S crystals is estimated to be about twice higher than those of GaAs in the same conditions. Thus, it is shown that gallium selenide nonlinear crystals are a promising material for the manufacturing of future-generation telecommunication devices operating at the wavelengths of 1.5 μm.

Acknowledgments

The studies were supported by the Ministry of Science and Higher Education of the Russian Federation, project #FSUS-2020-0029. The authors express their gratitude to the Shared Equipment Center "Spectroscopy and Optics" of the IA&E SB RAS for the provided instrumental support.

References

[1] Jiang, Biqiang, et al. 2020 High-efficiency second-order nonlinear processes in an optical microfibre assisted by few-layer GaSe *Light: Science & Applications* **9(1),** 1-8

[2] Ku, Shin An, et al. 2012 Optimal Te-doping in GaSe for non-linear applications *Optics express* **20(5),** 5029-5037.

[3] Nazarov, Maxim M, et al. 2012 Terahertz generation in $GaSe_{0.71}S_{0.29}$ and GaSe crystals via *eee*- and *eoo*-type optical rectification *37th International Conference on Infrared, Millimeter, and Terahertz Waves. IEEE*

[4] Kübler C, et al. 2004 Ultrabroadband detection of multi-terahertz field transients with GaSe electro-optic sensors: Approaching the near infrared *Applied physics letters* **85(16)** 3360-3362

[5] Yamashita, Masatsugu, et al. 2014 Ultra-broadband terahertz time-domain ellipsometric spectroscopy utilizing GaP and GaSe emitters and an epitaxial layer transferred photoconductive detector *Applied Physics Letters* **104(5)** 051103.

[6] Shi, Wei, et al. 2002 Efficient, tunable, and coherent 0.18–5.27-THz source based on GaSe crystal *Optics letters* **27(16)** 1454-1456

[7] Shi, Wei, et al. 2006 Observation of difference-frequency generation by mixing of terahertz and near-infrared laser beams in a GaSe crystal *Applied physics letters* **88(10)** 101101.

[8] Yan, Dexian, et al. 2017 High-average-power, high-repetition-rate tunable terahertz difference frequency generation with GaSe crystal pumped by 2 μm dual-wavelength intracavity KTP optical parametric oscillator *Photonics Research* **5(2)** 82-87.

[9] Rao, Zhiming, Xinbing Wang, and Yanzhao Lu. 2011 Tunable terahertz generation from one CO_2 laser in a GaSe crystal *Optics Communications* **284(23)** 5472-5474

[10] Tochitsky, Sergei Ya, et al. 2007 High-power tunable, 0.5-3 THz radiation source based on nonlinear difference frequency mixing of CO_2 laser lines *JOSA B* **24(9)** 2509-2516

[11] Nahata, Ajay, Aniruddha S. Weling, and Tony F. Heinz. 1996 A wideband coherent terahertz spectroscopy system using optical rectification and electro-optic sampling *Applied physics letters* **69(16)** 2321-2323

[12] Rybak, Alina, et al. 2021 Terahertz Optical Properties of $KTiOPO_4$ Crystal in the Temperature Range of (− 192)–150° C. *Crystals* **11(2)** 125

[13] Wang, Chun-Rui, et al. 2019 Phase-matching in KTP crystal for THz wave generation at room temperature and 81 K. *Infrared Physics & Technology* **97** 1-5

Tunable mid-infrared laser sources for trace-gas analysis

D B Kolker[1,2,3], I V Sherstov[1,2], A A Boyko[1,2,4], B N Nyushkov[1,3], E Y Erushin[1,2,3], N Y Kostyukova[1,2,3], A I Akhmathanov[5], A Y Kiryakova[3] and A V Pavluck[1]

[1] Novosibirsk State University, 1 Pirogova str., Novosibirsk, 630090, Russia
[2] Institute of Laser Physics SB RAS, 15B, Ac. Lavrentyev ave, Novosibirsk, 630090, Russia
[3] Novosibirsk State Technical University, 20 K. Marx ave, Novosibirsk, 630073, Russia
[4] Tomsk State University, 36 Lenin av. Tomsk, 634050, Russia
[5] Ural Federal University, 19 Mira Street, Yekaterinburg, 620002, Russia

Abstract. We demonstrate advanced experimental approaches to photoacoustic gas detection with tunable mid-infrared (mid-IR) laser sources of different types. A gas analyzer for registration of various gas components based on a tunable narrow-linewidth optical parametric oscillator (OPO) was designed and investigated. Using this OPO, the possibility of measuring the trace concentration (~2÷3 ppm) of methane (CH_4) in air was experimentally shown. The gas detection capability was enhanced by introducing injection seeding into the OPO. Another gas analyzer was based on a quantum cascade laser (tunable within the range ~7.6 ÷7.7 μm) and a resonant differential photoacoustic detector. Detection of the ultra-low concentration (~0.3 ppm) of methane in air was achieved (the standard dispersion was (1σ) ≈ (10–11) ppb with an integration time of 10 s). We compare the presented approaches and outline their further development.

1. Introduction

Tunable optical parametric oscillators, as well as quantum-cascade lasers, are currently used as laser sources for remote and local gas analysis. This paper presents experimental gas analytical systems and prototypes of operating devices based on optical parametric oscillators and quantum cascade lasers.

The mid-IR range (2-14 μm) is of great interest for various applications. It contains fundamental vibrational-rotational absorption bands of various gases, which allows remote [1, 2] or local gas analysis of the atmosphere [3, 4] using lasers with appropriate wavelengths. The absorption of water vapour in the atmosphere significantly limits the spectral regions in which laser radiation passes through the open atmosphere. The main transparency windows of the atmosphere are located in the ranges of 3-5 μm and 8-12 μm. In addition to water vapour, atmospheric carbon dioxide, which has strong broad absorption bands centred near 4.3 and 14 μm, also restricts the propagation of laser radiation. In the spectral range of 8-12 μm, tunable CO_2 lasers are traditionally used for various purposes (including atmospheric gas analysis). These lasers emit at various vibrational-rotational transitions in the range of 9.2-10.8 μm [5,6]. Optical parametric oscillators (OPOs) and quantum cascade lasers (QCLs) are widely used in the range of 3-5 μm and 8-12 μm [10]. Currently, gaseous pollutants such as sulfur dioxide (SO_2), nitrogen oxides (NO_x), and hydrocarbons (HC), as well as their reaction products, such as acids and oxidants, are of primary interest. SO_2, NO_x and hydrocarbons are the primary pollutants, while the reaction products are referred to as secondary pollutants. In industrialized countries with high traffic density, air pollution is mainly caused by human activities, that is, anthropogenic sources. In contrast, the contribution of natural sources to air pollution is often negligible. In Europe, the importance of SO_2 as an air pollutant has been slowly declining since ~ 1970 due to the decline in the sulfur content of fuel oil used for heating. On the contrary, NO_x and HC emissions are increasing. As a result, by the beginning of the XXI century, their concentrations turned out to be about 10 times higher than in the middle of the XX century. This increase is mainly related to cars, as up to 70% of NO_x emissions and ~ 25% of HC emissions are caused by vehicles.

The processes of global warming on Earth are also causing great concern to scientists around the planet. Currently, the background concentration of CO_2 as the main "greenhouse gas" has exceeded 400 ppm. Various international agreements (Kyoto Protocol, Paris Climate Agreement, etc.) call for limiting and reducing carbon dioxide emissions into the atmosphere in the future. For this, new technologies are being developed, the so-called. "Green energy" with renewable energy sources (solar energy, the use of wind energy, tidal power plants, etc.).

2. Optical parametric oscillators

2.1. *Fan-out-PPLN-based pulsed OPO with injection seeding*

OPOs are universal solid-state coherent sources with the possibility of wavelength tuning in the aforenamed spectral range. However, conventional OPOs without special measures for spectral narrowing feature relatively broadband radiation spectra (up to dozens of nanometers) which impede performing high-resolution spectroscopy. Therefore, special methods are applied in OPO design to narrower optical spectra of generated mid-IR radiation. These methods may include, for example, the use of injection-seed lasers [7-15], or the use of a spectral-selective element, such as an etalon, diffractive grating, or a volume Bragg grating, in the OPO cavity [16-19]. Here, we present an original narrow-linewidth pulsed OPO configuration which employs fan-out periodically-poled lithium niobate (fan-out PPLN) and injection seeding laser sources of two different types adapted for the PA sensing of methane. The developed experimental setup is shown in figure 1.

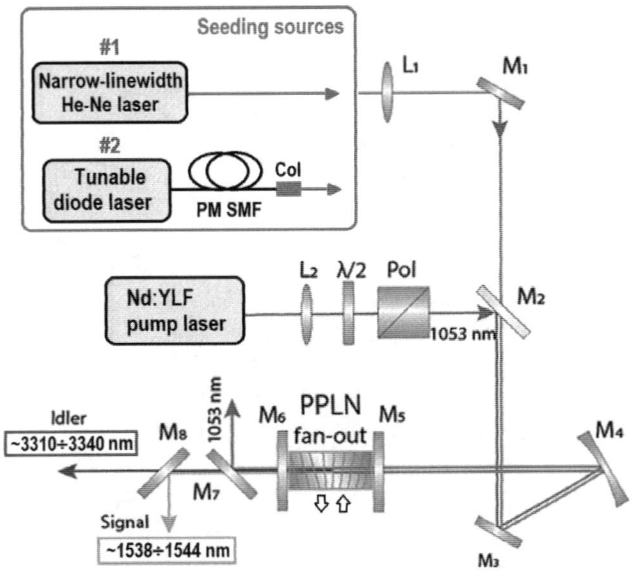

Figure 1. Schematic of the injection-seeded fan-out-PPLN-based OPO developed for PA gas sensing

Two alternative continuous-wave seed lasers were used in our experiments: a fibre-coupled distributed-feedback (DFB) diode laser with a megahertz linewidth at ~1.54 μm and a He-Ne laser with a kilohertz linewidth at 3.39 μm. The pumping and seeding radiation beams were matched with the OPO cavity by using lenses and dichroic mirrors. When the diode laser was used instead of, the He-Ne laser, the mirrors M_2 and M_5 were replaced accordingly with the seed wavelength. The OPO cavity was designed to be resonant for the signal wave at around 1.54 μm, thereby allowing the idler wave to be obtained at around 3.39 μm which allows for matching with an absorption line of methane. Without injection seeding laser, the tuning range of the idler wave can be in principle as wide as 1778 cm^{-1} (2.5-4.5 μm) [20]. The dichroic mirror M_7 filters out the residual pump radiation from the output beam containing signal and idler waves. In turn, the dichroic mirror M_8 splits the idler and signal waves, thereby allowing the idler wave to be used for PA gas sensing and the signal wave to be used for monitoring the OPO output with a commercial λ-meter (Angstrom IR II WS6). Thus, the developed OPO system can operate alternatively in two modes: with the injection seeding wave coinciding either with the signal wave (the diode seed laser at 1.54 μm) or the idler wave (the He-Ne seed laser at 3.39 μm). Both types of seeding provided spectral narrowing of the OPO system and thus improved the sensitivity of methane detection with a PA method. The options provided by these different types of seeding are the following: the DFB diode laser enables fine wavelength tunability [21], while the He-Ne laser provides stronger spectral narrowing [22] and can in principle ensure excellent wavelength stability of a metrological grade [23].

Figure 4 illustrates spectral narrowing achieved in the OPO upon injection of seeding radiation from the He-Ne laser. This spectral narrowing provides an improved response (higher sensitivity) of photoacoustic detection of the methane as illustrated in figure 2. This figure shows the response of the photoacoustic detector PAD-90 [24] to 3 different events: track 1 is the detector response to room air (background level of methane concentration); track 2 is the response to the test gas mixture N_2+1000 ppm CH_4 when seed laser is off; track 3 is the response to the test gas mixture N_2+1000 ppm CH_4 when seed laser is on. The PAD response is increased by a factor of ~2 when seed laser is used.

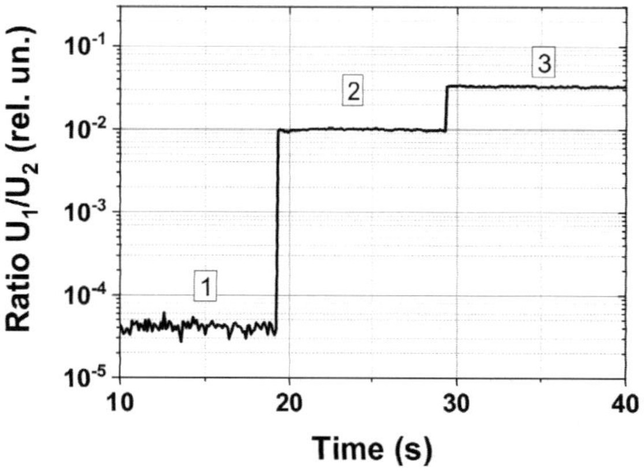

Figure 2. The response of the photoacoustic methane detector employing the developed OPO: track 1 is the detector response to room air (background level of methane concentration); track 2 is the response to the test gas mixture N_2+1000 ppm CH_4 when seed laser is off; track 3 is the response to the test gas mixture N_2+1000 ppm CH_4 when seed laser is on

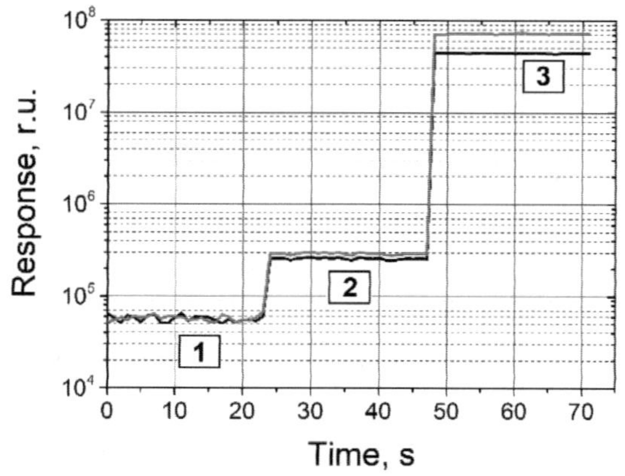

Figure 3. The responses of the photoacoustic methane detector employing the developed OPO under free-running (black trace) and injection-seeded (red trace) conditions: 1 – response to the dry nitrogen, 2 – response to the room air (~1.86 ppm of CH_4), 3 – response to the reference gas mixture (N_2+1000 ppm of CH_4)

We have also explored the possibility to enhance methane detection capabilities of the fan-out-PPLN-based OPO by combining its own wavelength tunability with the tunability of the aforenamed

Journal of Physics: Conference Series **2067** (2021) 012013 doi:10.1088/1742-6596/2067/1/012013

diode-laser seed source. Such injection seeding at the signal wavelength also ensures spectral narrowing of the OPO radiation and improves the response of the photoacoustic methane detector [21].

Figure 5 illustrates spectral narrowing achieved in the OPO upon injection of tunable seeding radiation from the DFB diode laser. This spectral narrowing along with fine wavelength tuning ensures improved response (higher sensitivity) of photoacoustic detection of the methane as illustrated in figure 3. This figure shows the response of the photoacoustic detector PAD-90 [24] to 3 different events under free-running (black trace) and injection-seeded (red trace) conditions: 1 – response to the dry nitrogen, 2 – response to the room air (~1.86 ppm of CH_4), 3 – response to the reference gas mixture (N_2+1000 ppm of CH_4). The use of the seed laser provides an almost double (~80%) increase in the photoacoustic response.

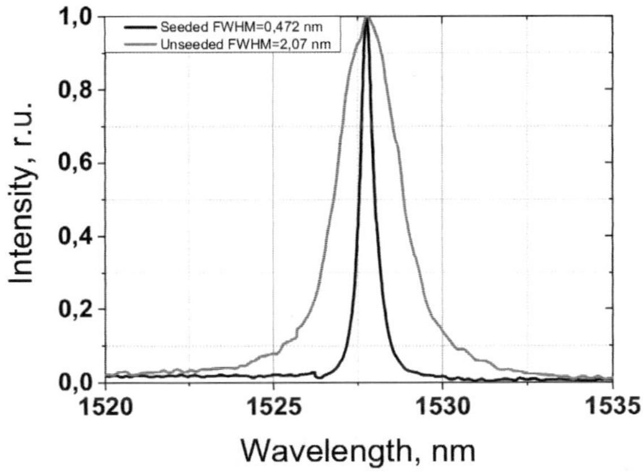

Figure 4. Optical spectra of the OPO signal radiation obtained without (red curve) and with (black curve) seeding radiation from the He-Ne laser

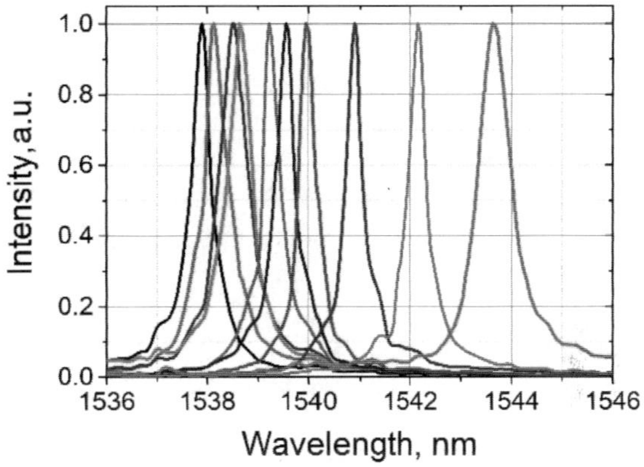

Figure 5. Optical spectra of the OPO signal radiation obtained with tunable seeding radiation from the DFB diode laser

It is worth noting that that the implemented injection seeding provides not only a significant narrowing of the OPO radiation spectrum but also noticeably stabilize its central wavelength [21].

We also anticipate that the above injection seeding technique can be further improved by applying a seed laser source with a much wider wavelength tuning range (like an ultra-widely tunable Er-fiber laser [25]). Useful options might be granted to the OPO also by a pulsed seed source with the arbitrary controllable timing (like that of a hybrid laser demonstrated in [26]). Such improvements will extend the tunability of spectral and temporal characteristics of the pulsed OPO radiation.

2.2. HGS-crystal-based OPO

Figure 6 shows an OPO experimental setup at 4.5-10 μm based on mercury thiogallate crystals. The OPO is pumped by an Nd: YLF laser in the region of 1.053 μm. The pulse repetition rate varied from 0.02 kHz to 4 kHz, and pulse duration was 5-7 ns. The pump radiation passes through the Faraday isolator, the λ/2 plate and, using the M_3 mirror and the M_4 dichroic mirror (passes the signal and idler, reflects the pump), is directed into the OPO cavity, which is formed by the M_1 and M_2 mirrors. The OPO cavity is formed by two flat mirrors: a metal mirror with a reflection coefficient of R = 97% for emission of all three waves (pump, signal, and idler) and a ZnSe input/output mirror with a high transmittance for pump radiation (T = 92% @ 1.053 μm) and idler wave (T = 80% @ 4.2-10.8 μm), and high reflectivity for signal wave emission (R ~ 95% @ 1.17-1.4 μm).

Journal of Physics: Conference Series **2067** (2021) 012013 doi:10.1088/1742-6596/2067/1/012013

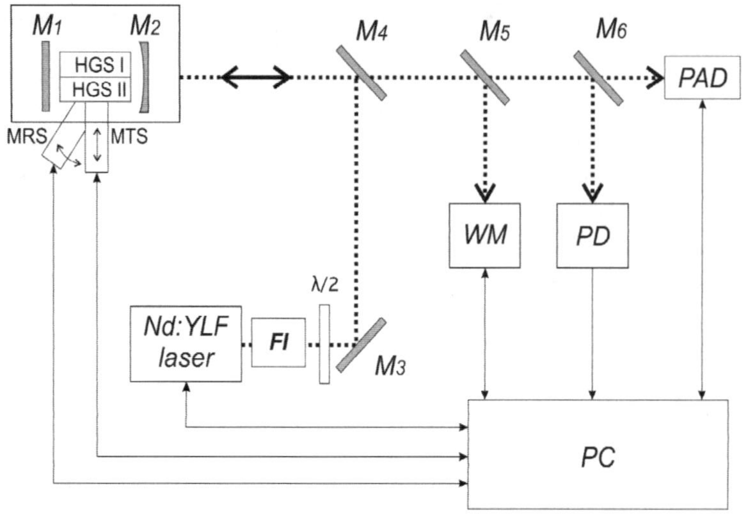

Figure 6. The part (4.5-10.8 μm) of combined OPO based on two $HgGa_2S_4$ crystals [9]

Both HGS crystals (Figure 6) are oriented for the second type of interaction (eo-e), but with different cut angles. The HGS I crystal was cut at an angle θ=60 °, φ=0 °, which corresponds to an idler wavelength of ~ 4.7 μm at normal incidence of pump radiation on the crystal, the HGS II crystal was cut at an angle θ=47 °, φ=0 ° (for an idler wavelength ~ 7.45 μm at normal incidence). The HGS I crystal has the following dimensions: 5x4.5x11 mm, and the HGS II 6x4.3x13.7 mm. To minimize optical losses in the cavity, a single-layer antireflection coating with a centre of 1.2 μm was applied to the ends of both crystals. The tuning of the OPO wavelength was carried out by motorized rotary stage (MRS) rotating the crystals relative to the axis of the OPO resonator, and switching between the crystals was carried out by motorized translation stage (MTS). Part of the OPO radiation on the signal wave is directed to the laser wavelength meter (WM), and the idler wave is directed to the optical-acoustic detector. In this case, a small part (4%) of the idler wave is directed by the M_6 mirror to the reference pyroelectric detector.

Figure 7 shows the OPO performance (idler wave) at ~5.05 μm. The HGS-OPO threshold was ~33 mJ/cm^2. The conversion efficiency of the pump energy into idler wave energy was about 3.1%, the differential efficiency was 4.28%, and the quantum efficiency was 14.86%.

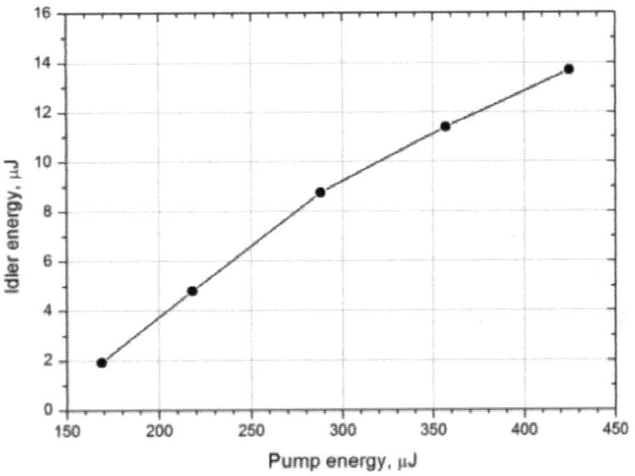

Figure 7. Performance of HGS-OPO at 5.05 µm

3. Quantum cascade lasers

In this work, we used a QD7500CM1 QCL (Thorlabs, Inc.), which is a compact semiconductor laser in the mid-IR range (λ ~ 7.5 ... 7.7 µm). This QCL is a so-called "DFB laser" - a distributed feedback laser that provides single-mode operations.

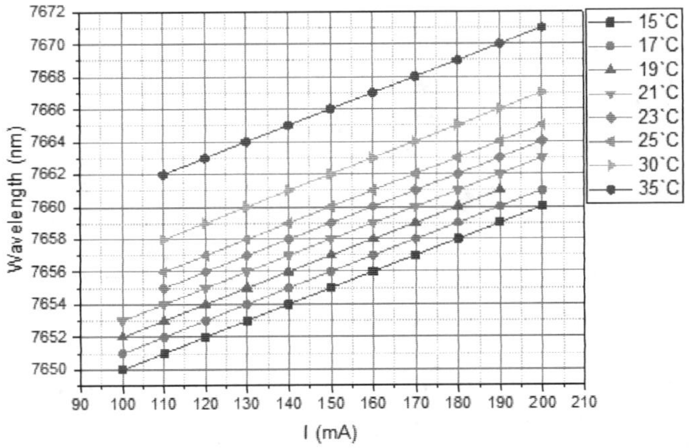

Figure 8. The QCL wavelength tuning curves. Wavelength tuning from injection current and temperature

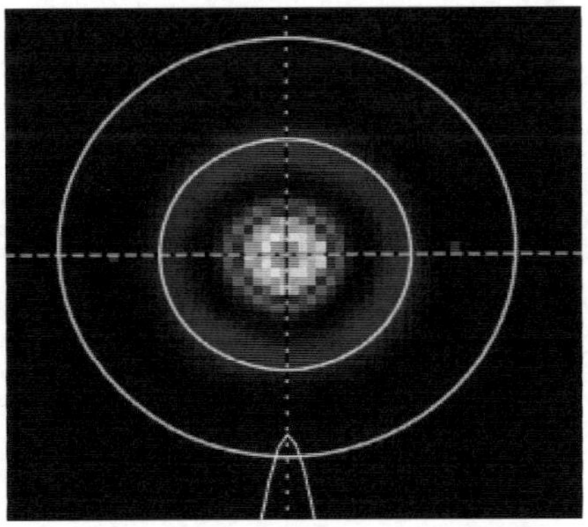

Figure 9. The QCL beam profile (Pirocam IV)

Figure 8 illustrates wavelength tuning of QD7500CM1 QCL (Thorlabs, Inc.) laser as a function of injection current with temperature as a parameter. From figure 8 we are estimating the QCL wavelength tuning range (injection current and temperature) as 7660±10 nm. At least we were obtained ~ 24 mW average powers when the injection current was swiped at the frequency 1700 Hz from 90 to 180 mA. Figure 9 shows the beam profile measured by Pirocam IV.

The photo-acoustic (PA) methane gas analyzer based on a quantum cascade laser (QCL; ~7.7 µm / 1800 Hz / 24 mW), a resonant differential PA detector, and a sealed gas-filled cell was investigated [20]. The measurement of methane concentration below the background value in the air (~0.3 ppm CH_4) is shown, the standard dispersion was (1σ) ≈ (10–11) ppb CH_4 with an integration time of 10 s. Under conditions of temperature instability (or emission wavelength) of QCL when normalized to a gas-filled cell, the relative measurement error of the CH_4 concentration does not exceed 3% (figure 10).

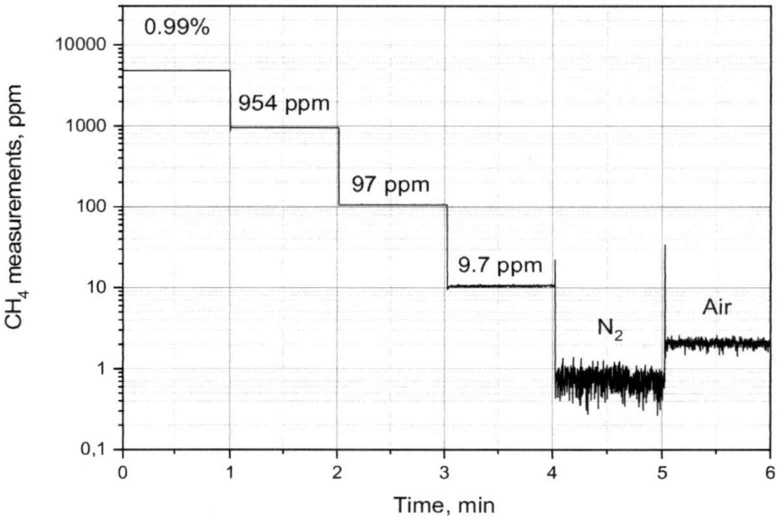

Figure 10. QCL-PAD methane gas analyzer. PAD response from different CH_4 concentrations [28]

4. Summary

Test-bed configurations of gas analyzers for registration of various gas components based on widely tunable OPOs operating in the mid-IR range (at wavelength from ~3 μm to ~10 μm) were demonstrated. The investigation of photoacoustic gas detection through the developed OPO was carried out using the example of methane. The OPO operated at an idler wavelength of ~ 3.3 μm which was used for the methane detection. The possibility of measuring the background concentration of methane in the air (~ 2–3 ppm CH_4) has been experimentally shown. The threshold sensitivity of the studied gas analyzer was (1σ) ~ 49 ppb CH_4.

The demonstrated spectral narrowing (and wavelength stabilization) of the OPO radiation utilizing injection seeding not only improves the response of the OPO-based photoacoustic gas detectors but also enhances the spectroscopic capabilities of OPOs in general. Owing to comparatively high power spectral density and wide tunability, such OPOs may successfully compete with other sources used for precise spectroscopy in mid-IR [28] and far-IR [29] ranges.

Besides that, a unique methane gas analyzer based on a quantum cascade laser at 7.65 μm has been developed as well. The device will be used to map the area for methane leaks and search for oil fields. The sensitivity is at the background level (1.86 ppm). The weight is 3 kg.

A decrease in the average QCL radiation power (~24-12-6 mW) leads to a noticeable deterioration in the threshold sensitivity of the PA gas analyzer. It is recommended to increase the average QCL power level to ~50 mW. The dynamic range of measuring the concentration of methane of the PA gas analyzer in the linear mode was ~4 decades (from 0.3 ppm to 2000-3000 ppm CH_4)

Acknowledgments

This work was partially supported by the Ministry of Science and Higher Education of the Russian Federation (grant FSUS-2020-0036). The Fan-out MgO: PPLN chip was prepared by A.A. and V.S with support from Russian Foundation for Basic Research (grant RFBR-mk-18-29-20077). The CH_4 spectroscopic experiments were funded by RFBR, project number 19-32-60055.

Journal of Physics: Conference Series **2067** (2021) 012013 doi:10.1088/1742-6596/2067/1/012013

References

[1] Balandin S F and Shishig S A 2006 *Proc. SPIE* 6160 (Tomsk) **Vol. 6160** 61601Z
[2] Ivashenko M V and Sherstov I V 2000 *Quantum Elec.* **30** 747-52
[3] Philippe Repond and Markus W Sigrist, *Appl. Opt.* **35** 4065
[4] Sherstov I V, Kapitanov V A, Ageev B G, Karapuzikov A I and Ponomarev Y N 2004 *Atmospheric Ocean. Opt*. **17** 119-23
[5] Karapuzikov A I, Malov A N and Sherstov I V 2000 *Infrared Phys. Technol.* **41** 77-85
[6] Sherstov I V, Buchkov K V, Vasilyev V A, Karapuzikov A I, Spitsyn V V and Chernikov S B 2005 *Quantum Elec.* **18** 270
[7] Fix A, Schröder T, Wallenstein R, Haub J G, Johnson M J and Orr B J 1993 *J. Opt. Soc. Am. B* **10** 1744
[8] Fink T, Büscher S, Gäbler R, Yu Q, Dax A and Urban W 1996 *Rev. Sci. Instrum.* **67** 4000-04
[9] Petrov V 2015 *Prog. Quantum. Electron.* **42**. 1-106
[10] Elder I, Legge D, Beedell J and Marchington R 2006 *Technical Digest Advanced Solid State Photonics*. paper MB20
[11] Mahnke P, Klingenberg H H, Pix A and Wirth M 2007 *Appl. Phys. B* **89** 1–7
[12] White R T, He Yabai, Orr B J, Kono M and Baldwin K G H 2003 **28** 1248-50
[13] Plusquellic D F, Votava O and Nesbit D J 1996 *Appl. Opt.* **35** 1464-72
[14] Chen X, Zhu X, Li S, Ma X, Zhang J, Liu J and Chen W 2019 *Proc. SPIE (Hangzhou, China)* **11185** 111851C
[15] Takidaa Y, Nawata K and Minamide H 2020 *APL Photonics* **5** 061301
[16] Jacobsson B, Tiihonen M, Pasiskevicius V and Laurell F 2005 *Opt. Lett.* **30** 2281-83
[17] Henriksson M Doctoral thesis *Tandem optical parametric oscillators using volume Bragg grating spectral control,* 2010, 73 pages.
[18] Henriksson M, Tiihonen M, Pasiskevicius V and Laurell F 2007 *Appl. Phys. B* **88** 37–41
[19] Peng Y, Wei X, Nie Z, Luo X, Peng J, Wang Y and Shen D 2015 *Opt. Express* **23** 30827-32
[20] Kolker D B, Sherstov I V, Pustovalova R V, Kostyukova N Y, Boyko A A and Zenov K G 2017 *Quantum Elec.* **47** 14-19
[21] Erushin E, Nyushkov B, Ivanenko A, Akhmathanov A, Shur V, Boyko A, Kostyukova N and Kolker D 2021 to be published at *Laser Phys. Lett.*
[22] Boyko A A, Erushin E Y, Kostyukova N Y, Miroshnichenko I B, Kolker D B 2021 *Instrum. Exp. Tech.* **64** 254–58
[23] Belfi J, Beverini N, Bosi F, Carelli G, Di Virgilio A, Kolker D, Maccioni E, Ortolan A, Passaquieti R and Stefani F 2012 *J. Seismol.* **16** 757 – 66
[24] Sherstov I V and Kolker D B 2020 *Quantum Elec.* **50** 1063–67
[25] Nyushkov B, Kobtsev S, Antropov A, Kolker D and Pivtsov V 2019 *J. Light. Technol.* **37** 1359-63
[26] Nyushkov B, Ivanenko A, Smirnov S, Shtyrina O and Kobtsev S 2020 *Opt. Express*. **28** 14922-32
[27] Sherstov I V, Kolker D B, Boyko A A, Vasiliev V A and Pustovalova R V 2021 *Infrared Phys. Technol.* **117** 103858
[28] Vainio M and Karhu J 2017 *Opt. Express* **25** 4190-200
[29] Beverini N, Carelli G, De Michele A, Maccioni E, Nyushkov B, Sorrentino F and Moretti A 2005 *Opt. Lett.* **30** 32-34

Study of Ne:He Plasma of a Periodically Pulsed Discharge for Optically Pumped Rare Gas Laser

A P Torbin, A K Chernyshov, M I Svistun and P A Mikheyev

Lebedev Physical Institute: Novo-Sadovaya st.221, 443011, Samara, Russia

torbinalex@gmail.com

Abstract. Optically pumped rare gas lasers (OPRGL) suggested recently as a chemically inert analog of diode-pumped alkali lasers are under extensive study at present. OPRGLs employ metastable atoms of heavier rare gases (Rg*) in He bath produced in discharge plasma. Ar* OPRGL is the most popular system at present, due to presence of a narrow band diode pump and abundance of Ar. However, Ne* OPRGL is interesting due to its visible lasing wavelength at 703.2 nm nm and presence of channels of energy transfer in Ne-He plasma that facilitate Ne* production. We present the first results of experiments with Ne* OPRGL that include Ne* number density in its active medium, and lasing experiments to determine pumping threshold for s5 → p9 transition in a transverse pumping configuration using a narrow band pulsed dye laser as a pump.

1. Introduction

Hybrid gas-phase / solid-state lasers are promising for the development of highly efficient high-power laser systems with good beam quality. The mainstream solution at present is the diode-pumped alkali metal vapor laser (DPAL), but its further development is limited by a number of problems: chemically aggressive nature of alkali metals; the need for heating of the laser cell; arrangements for circulation of highly reactive gaseous medium; the use of methane or ethane as a buffer gas which are fire hazardous and chemically active in the conditions of DPAL. In this regard, the development of a chemically inert analogue of DPAL is an urgent scientific and technical problem. If we restrict ourselves to the search for alternative systems pumped from the ground state, the choice turns out to be limited. However, over the past few years, the concept of optical pumping of atoms from metastable electronic states has been developed. The production of metastable atoms can be achieved using a low-power electric discharge.

Optically pumped rare gas lasers (OPRGL) are under extensive study at present. OPRGLs employ metastable atoms of heavier rare gases (Rg*) in He bath produced in discharge plasma. It is well known that Rg excited to metastable states are similar to alkali metal atoms [1]. Amplification and lasing were observed during optical pumping of metastable atoms of heavy inert gases Rg* = Ne*, Ar*, Kr* and Xe* [2–4]. OPRGLs are, in essence, three-level systems. The typical energy level diagram of OPRGL with lasing and pump wavelengths is presented in figure 1. Optical pumping is achieved by excitation of the $s_5 \rightarrow p_9$ transition (here and after the Paschen notation is used). A population inversion between p_{10} and the s_5 level is generated by rapid collisional relaxation from p_9 to p_{10}. The collisional partner for this step is He. The system then lases on the $p_{10} \rightarrow s_5$ transition [5]. Hence, the entire lasing medium is chemically inert. Analogous to DPAL, this lasing medium can

Journal of Physics: Conference Series **2067** (2021) 012014 doi:10.1088/1742-6596/2067/1/012014

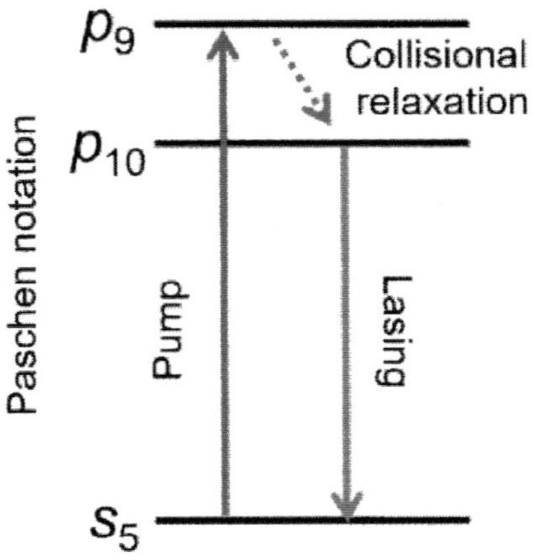

Rg*-He

	λ_{pump}, nm	λ_{lase}, nm
Ne*	**640.2**	**703.2**
Ar*	811.5	912.3
Kr*	811.3	893.1
Xe*	881.9	980.0

Figure 1. Typical energy level diagram of OPRGL.

provide hundreds of watts of laser output out of cubic centimeter. Diode-pumped rare gas laser systems may ultimately outperform the classic DPAL system. Recent progress includes CW laser demonstrations of Ar* and Xe* lasers operating at 7.3 and 1.3 W, respectively [6,7]. However, Ne* OPRGL is interesting due to its visible lasing wavelength (703.2 nm) and presence of channels of energy transfer in Ne-He plasma that facilitate Ne* production.

The aim of this work was to obtain lasing in a Ne*/He mixture in a pulsed regime with transverse pumping, to study the effect of the active medium parameters on lasing and to measure the pumping threshold. The studies performed are necessary to verify the numerical model of the OPRGL.

2. Experiment

Schematic view of the experimental setup for the experiments with lasing is shown in figure 2. Metastable neon atoms Ne* were produced in the plasma of a repetitievly pulsed glow discharge (RPD). The discharge was ignited between a pair of tungsten (cathode) and platinum-coated titanium electrodes with geometric dimensions of 1.5 × 0.4 cm and 3 mm interelectrode distance. A photograph of electrodes with the burning discharge is shown in figure 3. The electrodes were located inside a 0.5 L stainless steel chamber (figure 4) with mutually perpendicular channels for optical pumping and laser output.

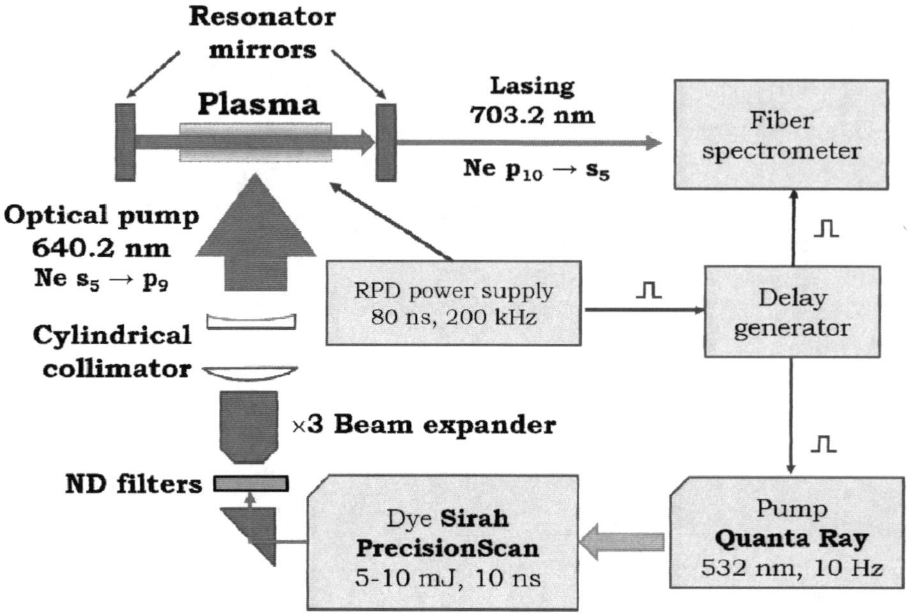

Figure 2. Schematics of the experimental setup for the experiments with lasing.

Figure 3. Stainless steel discharge chamber.

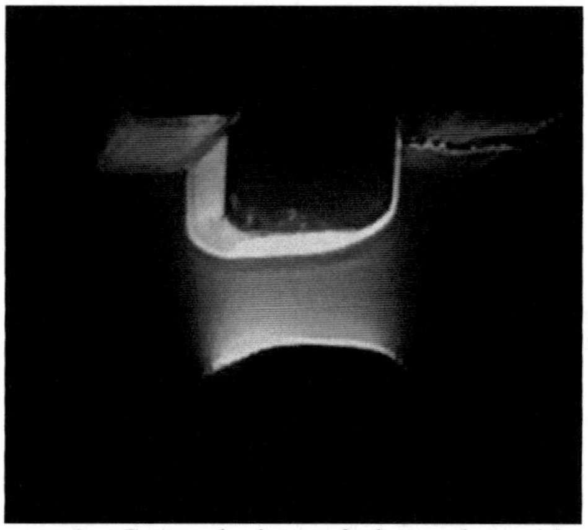

Figure 4. General view of electrodes with the discharge plasma (camera exposure 1/1000, ISO 100).

The RPD could exist in a wide range of pressures - from several Torr to more than atmospheric pressure. A laboratory made high-voltage pulse generator operated at 200 kHz with pulse duration 80 ns to sustain the discharge. Obtaining the required number densities of metastable atoms requires a high purity of the discharge chamber and gases, therefore, the experimental setup was equipped with oil-free pumping, chemically inert tubing and Swagelok vacuum fittings. A stable gas mixture flowrate (2.3 standard liters per minute) through the discharge chamber was organized with the help of gas flow controllers Bronkhorst F-201CV.

Transverse optical pumping at 640.2 nm, corresponding to the Ne $s_5 \rightarrow p_9$ transition, was provided by a dye laser Sirah PrecisionScan, pumped by the second harmonic of a Quanta-Ray Pro Nd:YAG laser. Pump laser radiation was formed into a narrow oval beam to provide a high power density of optical pumping with the help of a x3 beam expander and a system of cylindrical lenses.

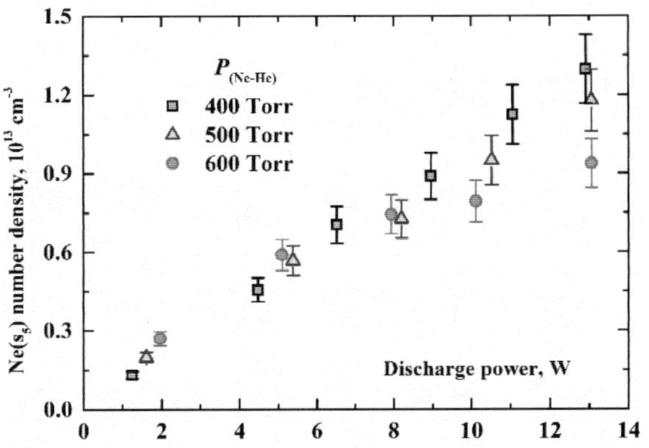

Figure 5. Ne(s_5) number densities dependencies on discharge power (4% Ne in He) for three total pressures.

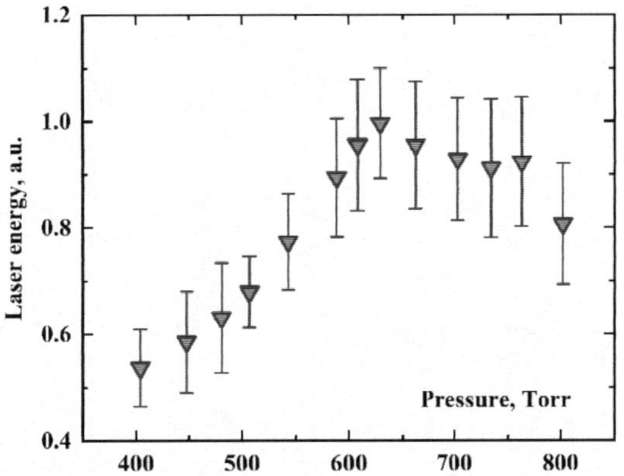

Figure 6. Lasing pulse energy against total pressure of Ne-He mixture: 4% Ne in He, discharge power 20 W, pump intensity 2 MW $\mathrm{cm^{-2}}$.

Then, the oval laser beam was cut using a 15 x 1 mm slit and transferred to the center between the electrodes. The input and output of radiation was carried out through quartz windows 38.1 mm in diameter and 3 mm thick, inclined at the Brewster angle. The pump intensity varied by no more than 20% along the optical axis of the resonator. The laser resonator consisted of two plain mirrors separated by ~ 40 cm. The output mirror transmission was specified by the manufacturer to be ~5%.The intracavity losses did not exceed 8%. Lasing at 703.2 nm, corresponding to the Ne $p_{10} \rightarrow s_5$ transition, was recorded with the help of an Avantes fiber spectrometer, the input fiber of which was located on the optical axis of the resonator.

In experiments determining the lasing threshold, the pump intensity was attenuated using Thorlabs' neutral filters. The data for the filters' transmission at 640.2 nm was obtained from the Thorlabs website. Pump laser energy was measured by a power meter Ophir PE50 BF-C.

A sync pulse from the power supply triggered a Stanford Research Systems DG645 delay generator, which triggered the pump laser with a predetermined delay of about 1 μs. Such a scheme made it possible to position the pump pulse at any place of the voltage period applied to the electrodes, thereby ensuring a constant number density of metastables for each pump pulse.

Number density measurements of Ne(s_5) metastable atoms in discharge zone were carried out using the tunable diode-laser absorption spectroscopy method. A tunable diode laser provided continuous probe radiation, which was recorded by a photodiode. The observed absorption profiles of the $s_5 \rightarrow p_9$ transition were fitted to a Voigt function. Further, the Ne(s_5) number densities were determined in the same manner as in [8]. The RPD produced [Ne(s_5)] ~ 10^{13} $\mathrm{cm^{-3}}$, resulting in opaqueness near the line centre. To avoid incorrect interpretation of the data recorded in the opaque range, the central part of the absorption profile was excluded from the fit.

3. Results and discussion

Figure 5 presents the results of the tunable diode-laser absorption spectroscopy experiments. Ne(s_5) number densities dependencies on discharge power were determined for three total pressures of Ne-He mixtures in the discharge chamber. The neon content in helium was 4%. The observed results demonstrate two experimental facts. First, number densities of Ne(s_5) are weakly dependent on gas pressure. Second, the electrode materials chosen is present study (tungsten cathode and platinum-coated titanium anode) and the RPD power supply system provided necessary for efficient lasing [Ne(s_5)] ~10^{13} $\mathrm{cm^{-3}}$.

Experiments with transverse optical pumping by a narrow band dye laser showed that lasing in the Ne*:He mixture could be produced easily. Pumping at the maximum of pump intensity distribution 2 MW $\mathrm{cm^{-2}}$ was transferred to the central region of the discharge at an equal distance from the

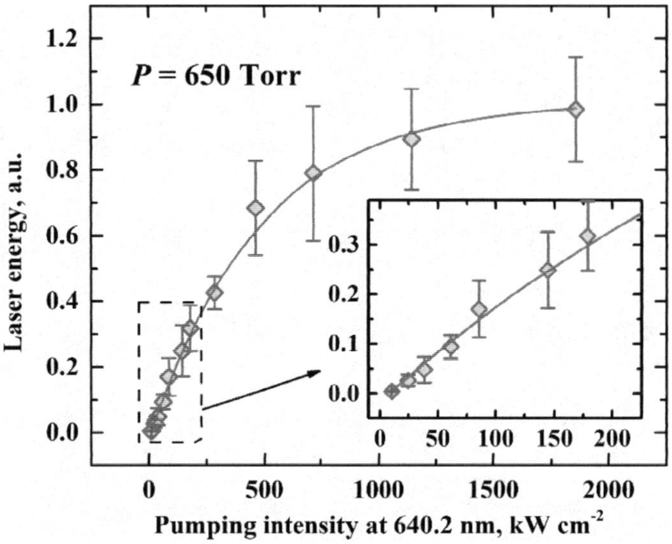

Figure 7. Laser pulse energy against pump intensity at 640.2 nm (Ne transition $s_5 \rightarrow p_9$): total pressure 650 Torr, 4% Ne in He, discharge power 20 W. Smooth line – the result of exponential approximation

electrodes. A preliminary experimental check of the effect of the neon content on the lasing efficiency showed that when 4% Ne in He was exceeded, the laser pulse energy ceased to grow. Figure 6 shows the dependence of the lasing energy on the total pressure in the discharge chamber, obtained at an electric discharge power 20 W, 4% Ne in He and pump intensity 2 MW cm^{-2}. The lasing reaches maximum near 600 Torr and then a slight decrease is observed, which, taking into account accuracy of the measurements, can be regarded as a plateau up to the atmospheric pressure. This data correlate well with our previous results obtained for the Ar*-He system [2] where a similar dependence was observed.

The main goal of this work was to determine the pump intensity threshold of the for $s_5 \rightarrow p_9$ transition. The 640.2 nm pump radiation was attenuated using calibrated Thorlabs neutral light filters. In the experiments, it was found that the lasing threshold strongly depended on various geometric factors, such as the width of the pump laser beam, the position of the pumping region between the electrodes and the quality of the resonator alignment, as well as on quality of synchronization of the discharge with the pump laser. Figure 7 shows the typical dependence of the lasing pulse energy on the pump intensity obtained at a pressure in the discharge chamber 650 Torr, 4% neon content in the gas mixture and a discharge power 20 W. The exponential approximation of the dependence section where weak lasing appears allowed determining the minimum value of pumping intensity threshold to be 9.4 ± 1.1 kW cm^{-2}. For comparison, according to the results of our previous work [2] in the same discharge chamber the measured pumping threshold for pumping of the p_9 state of argon in Ar*-He was 250 ± 150 W cm^{-2}.

4. Conclusion

In this work, lasing was obtained employing metastable neon atoms in a mixture with helium with transverse optical pumping at the Ne $s_5 \rightarrow p_9$ transition using a 200 kHz repetitively pulsed glow discharge as Ne* source and a pair of tungsten (cathode) and platinum-coated titanium electrodes. The pumping threshold for lasing in the Ne*/He medium was measured for the first time and amounted to 9.4 ± 1.1 kW cm^{-2}. It was shown that with pressure the energy of laser pulses reached maximum at near 600 Torr, with further pressure increase pressure dependence was weak. The number densities of metastable neon atoms Ne(s_5) were weakly dependent on the total pressure of the gas mixture. The obtained results will be used for design and development of Ne OPRGL.

References

[1] Setser D W, Dreiling T D, Brashears H C and Kolts J H 1979 Electronic excitation. Analogy

between electronically excited state atoms and alkali metal atoms *Faraday Discuss. Chem. Soc.* **67** 255–72

[2] Mikheyev P A, Chernyshov A K, Svistun M I, Ufimtsev N I, Kartamysheva O S, Heaven M C and Azyazov V N 2019 Transversely optically pumped Ar:He laser with a pulsed-periodic discharge *Opt. Express* **27** 38759

[3] Mikheyev P A, Han J, Clark A, Sanderson C and Heaven M C 2017 Production of Ar and Xe metastables in rare gas mixtures in a dielectric barrier discharge *J. Phys. D. Appl. Phys.* **50** 485203

[4] Rawlins W T, Galbally-Kinney K L, Davis S J, Hoskinson A R, Hopwood J A and Heaven M C 2015 Optically pumped microplasma rare gas laser *Opt. Express* **23** 4804–13

[5] Sun P, Zuo D, Mikheyev P A, Han J and Heaven M C 2019 Time-dependent simulations of a CW pumped, pulsed DC discharge Ar metastable laser system *Opt. Express* **27** 22289

[6] Han J, Heaven M C, Moran P J, Pitz G A, Guild E M, Sanderson C R and Hokr B 2017 Demonstration of a CW diode-pumped Ar metastable laser operating at 4 W *Opt. Lett.* **42** 4627

[7] Clark A B, Hokr B H, Sanderson C R, Ballmann C W, Han J, Xu K G and Heaven M C 2019 Demonstration of a quasi-CW diode-pumped metastable xenon laser *Opt. Express, Vol. 27, Issue 24, pp. 36011-36021* **27** 36011–21

[8] Hübner S, Sadeghi N, Carbone E A D and Van Der Mullen J J A M 2013 Density of atoms in Ar*(3p54s) states and gas temperatures in an argon surfatron plasma measured by tunable laser spectroscopy *J. Appl. Phys.* **113** 143306

Effect of terahertz radiation on intermolecular interactions of albumin under aerobic and anaerobic conditions

E A Nemova[1], G G Dultseva[2], N A Nikolaev[1,3], O P Cherkasova[1,4]

[1]Institute of Laser Physics SB RAS, 15 B Academician Lavrentiev Ave., Novosibirsk, Russia
[2]Voevodsky Institute of Chemical Kinetics and Combustion SB RAS, 3 Institutskaya St., Novosibirsk, Russia
[3]Institute of Automation and Electrometry SB RAS, 1 Academician Koptyug Ave., Novosibirsk, Russia
[4]Novosibirsk State Technical University, 20 Karl Marks Ave., Novosibirsk, Russia

Corresponding author: Nemova Evgenia, email: endy@ngs.ru

Abstract. The effect of THz radiation on coupling between albumin molecules under aerobic and anaerobic conditions is assessed through an EPR-quantified procedure. Rearrangements induced by terahertz radiation are found to affect the hydrogen bonding network causing an increase in the rate of bimolecular interactions. Molecular mechanism of the observed effect is considered involving the interactions between the functional groups of albumin molecules. Irradiation causes conformational changes in albumin molecules, which involves changes in the steric states of molecules. These rearrangements hinder or simplify the adsorption of specific agents on the reactive sites of albumin molecules.

1. Introduction

The appearance of contemporary THz sources and detectors stimulates a rapid development of medical applications and causes an increased exposure of human population to THz waves [1, 2]. At the same time, the biological effects of THz radiation remain insufficiently investigated [3-7].

Intermolecular interactions are crucial for the transport functions of albumins, so interactions with different agents to be transported *in vivo* are the subject of extensive investigation. Albumin conformations determine the possibilities of binding with oxygen, nitrous oxide, various metal complexes of biological significance. The effect of preliminary terahertz radiation on conformation changes in albumin molecules has already been traced as changes in albumin binding with molecular oxygen, in albumin-mediated NO - NO_2 equilibria, and in the rates of interaction with ozone [8]. The possibility to assess conformation changes in globular proteins is provided by the presence of paramagnetic centers in these biopolymers: the parameters of their EPR spectra (the general structure, hyperfine splitting constants, line width, and decay kinetics) allow one to determine chemical structure, mobility, oxidation state, lifetimes of paramagnetic centers formed and decayed in various interactions [9]. Paramagnetic centers permanently existing in albumin molecules in aqueous solutions interact not only with dissolved oxygen molecules and nitrous oxide but also with the same centers of other albumin molecules. These kinds of interactions are strongly dependent on conformation states, so it could be expected that preliminary irradiation of filmed albumin preparations in the terahertz range (before dissolution) would affect also intermolecular binding between protein molecules.

The goal of the work was to quantify the influence of terahertz radiation on coupling between albumin molecules in the presence of oxygen and in its absence. To assess the rates of intermolecular interactions, a quantitative EPR-based technique was applied in combination with high performance liquid chromatography and the methods of qualitative chemical analysis.

2. Experimental

Bovine serum albumin BSA (Sigma-Aldrich, lyophilized powder) was dissolved in distilled water to prepare a solution with the concentration 1 mg/mL. A drop of this solution (0.05 mL) was applied onto a substrate made of crystal quartz and dried in the air.

The sample was irradiated with terahertz waves using a THz-TDS described in detail in [10]. The radiation spectral range was 0.2–1.5 THz, in the pulse mode, with pulse duration about 2.5 ps and a repetition rate of 80 MHz. The estimated THz irradiance was ~10 mW/cm^2 and exposure time was 60 min. Then the samples were dissolved in deionized water (0.05 mL) for EPR investigation.

The solution was poured in a quartz flat cell for EPR spectroscopy of aqueous solutions. EPR spectra were recorded with a Bruker EMX spectrometer (Germany) within the X band, with amplification up to $5 \cdot 10^5$. Dihydropyrazine-1,4-dioxide (DPDO) was used as a spin probe precursor to examine paramagnetic centers in BSA. The actual spin probe was formed *in situ* as a result of the oxidation of this dinitrone compound.

Chromatographic studies were carried out with a microcolumn Milikhrom A-02 chromatograph equipped with an UV detector (Econova, Russia). Chromatograms were recorded in the direct gradient mode and in the ion pair chromatographic mode. The volume of BSA samples introduced into the chromatograph was 20 μL. Elution rate was 150 μL/min. Elution was carried out in the gradient mode: eluent A – acetonitrile (Sigma-Aldrich, for HPLC, gradient grade, ≥99.9 %,), eluent B – 2×10^{-2} M phosphate buffer (pH=3). The initial concentration of eluent B was 3 %, the gradient: 0-600 μL – 3 %, 600-1800 μL – 30 %, 1800-2500 μl – 50 %. The ion pair chromatographic mode was used to provide separation of coupled BSA molecules held together through amino groups.

Intermolecular interactions were simulated using the CS Chem 3D Pro software, version 5.0 (Cambridge Soft Corporation, UK).

3. Results and discussion

3.1. Interpretation of the EPR spectra. Mobility of weak paramagnetic centers

The effects of THz radiation detected by us previously included an increase in the number of paramagnetic centers, lower mobility of these centers and increased adsorption of molecular oxygen in THz-irradiated BSA samples in comparison with non-irradiated ones [8]. A weak broad triplet (a_N = 0.94 mT) observed in EPR spectra of the aqueous solutions in addition to rather intense signals (a_N = 1.47 mT) related to the paramagnetic centers in BSA was attributed to a nitrogen-to-nitrogen bound cluster, because the hyperfine splitting constant for nitrogen atom has changed dramatically. The intensity of this weak triplet was found to be affected by preliminary irradiation in the terahertz range. In addition to differences in hyperfine splitting constants, the width of lines in the EPR spectra recorded in the differential form was found to decrease. This means that THz irradiation causes an increase in the mobility of paramagnetic centers in BSA. This was rather unexpected result, since the mobility of paramagnetic centers characterized by the strong EPR signal was previously demonstrated to decrease under irradiation. A possible explanation of an increase in the mobility of weak paramagnetic centers under irradiation might be their dependence on oxygen adsorption, which was proved to be stronger in irradiated BSA sample.

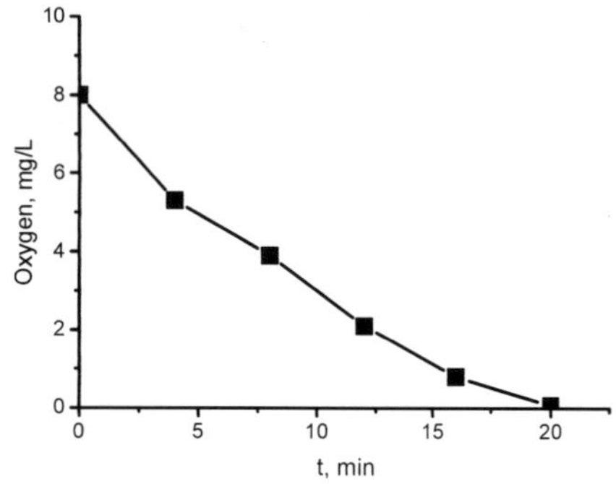

Figure 1. Oxygen concentration (mg/L) in solution depending on the time (min) of bubbling with argon.

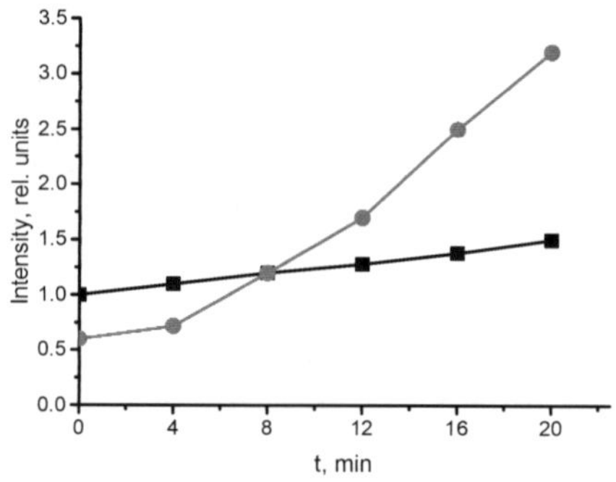

Figure 2. Intensity of the weak EPR signal in THz-irradiated (circles) and non-irradiated (squares) BSA film samples.

3.2. The dependence of the intensity of weak EPR signals on oxygen concentration in solution

To test the assumption concerning the interaction of BSA with oxygen dissolved in water, we studied the dependence of the weak EPR signals on irradiation in the absence of oxygen. Oxygen adsorption on BSA after film dissolution was found to be an equilibrium process, so oxygen was removed from solution by bubbling argon for 20 min. Oxygen concentration was measured by means of iodometric titration. The dependence of oxygen concentration C_{ox} in solution on the time t of bubbling with argon under our experimental conditions is presented in figure 1. This curve shows the values averaged over 10 independent measurements for each time point. The error did not exceed ±2 %. In subsequent measurements, oxygen concentration was estimated from t using the curve shown in figure 1.

After the removal of molecular oxygen, the integral intensity of these weak EPR signals was detected to increase by a factor of 2.4 in comparison with non-irradiated BSA samples treated in the same way. The dependence of the integral intensity of this weak EPR signal on t is shown in figure 2. To check whether these signals are due to the clusters formed by several BSA molecules interacting through the nitrogen atoms of their heterocycles or due to the interactions in which the spin probe DPDO was involved, chromatographic investigation was carried out.

3.3. Chromatographic investigation of the system BSA – DPDO: the effect of THz-irradiation

Chromatographic detection was performed by measuring the optical absorption at 230, 246 and 258 nm. An intense peak related to unbound DPDO was present in all chromatograms, though its intensity was about 20 % lower in the chromatograms of THz-irradiated BSA. Two very broad peaks with low intensity were detected in the chromatograms. Their absorption spectra were identical, but retention times differed greatly. Relying on concentration dependencies, we suppose that these peaks relate to monomeric BSA and a cluster formed by two BSA molecules held together through rather strong hydrogen bonding. Under aerobic conditions the concentration of clusters formed in non-irradiated BSA is slightly less than that in the THz-irradiated sample, while under anaerobic conditions (after removal of dissolved molecular oxygen) coupling becomes substantially stronger in irradiated BSA samples than in non-irradiated ones. Relative concentrations of clustered BSA molecules (dimers) in the aqueous solutions of THz-irradiated and non-irradiated BSA samples are shown in Figure 3.

3.4. Semi-empirical simulation of clustering in BSA solution

To determine the reasons of changes in intermolecular coupling of BSA, simulation was carried out to evaluate the strength of hydrogen bonds formed between different functional groups of all the

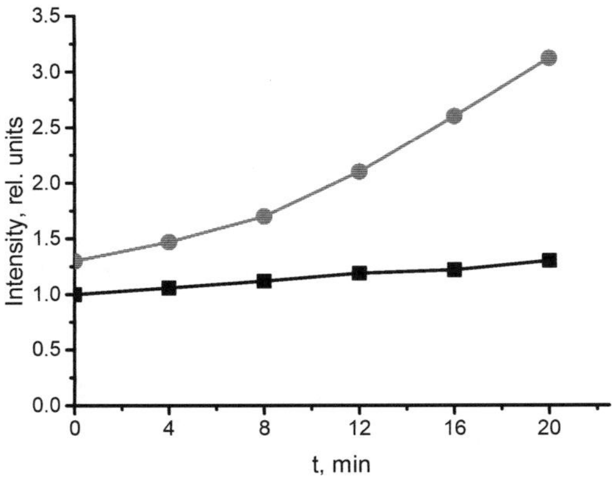

Figure 3. Relative concentrations of clustered BSA (dimers) in the aqueous solutions of THz-irradiated (circles) and non-irradiated (squares) BSA samples: chromatographic determination.

compounds present in the system. It was shown that oxygen molecules adsorbed on BSA molecules occupy the same positions that are involved in intermolecular coupling of BSA molecules. Simulation of the structure of a cluster formed by coupling revealed that in the presence of oxygen two BSA molecules are held together mainly through SH---O_2---HS bonds, while under anaerobic conditions intermolecular coupling is provided mainly by -S----S- interactions, with a small (about 15 %) contribution from the interactions of a nitrogen atom in the heterocycle of one BSA molecule with a hydroxyl group of another BSA molecule. So, conformational transition induced by irradiation within the THz range removes the steric hindrance for oxygen adsorption, and removal of oxygen from the solution makes these positions free for intermolecular coupling between BSA molecules. The studied effects of terahertz waves may lead to diverse biological consequences, for example, as those reported [11]: irradiation within the terahertz range affects cell adhesion, which may be a consequence of conformation changes in the structures on cell shells. A link is still to be traced from the molecular effects of terahertz radiation to its effects on much more complicated systems, such as cells, tissues, and organisms.

4. Conclusions

Irradiation of albumin, one of the representatives of biomacromolecules, within the terahertz range affects the rotational and low-frequency vibrational transitions in these molecules, thus inducing changes in their conformations. These changes affect the steric states, which simplifies or hinders the adsorption of specific agents. Under anaerobic conditions, these conformational changes enhance intermolecular coupling, which was demonstrated experimentally for albumin molecules. Conformational transition induced by irradiation within the THz range removes the steric hindrance for oxygen adsorption. The removal of oxygen from definite positions makes these positions free for intermolecular coupling between albumin molecules. The studied effects of terahertz waves may lead to diverse biological consequences, but exact mechanisms of the expected effects remain unknown and are to be studied.

Acknowledgments

The research is supported by Russian Ministry of Science and Higher Education within the State assignment (AAAA-A17-117030310290-5).

References

[1] Cherkasova O P, Serdyukov D S, Nemova E F et al 2021 *J. Biomed. Opt.* **26**(9), 090902.
[2] Cherkasova O P, Serdyukov D S, Ratushnyak A S et al 2020 *Optics & Spectroscopy* **128**(6) 855.
[3] Zaytsev K I, Dolganova I N, Chernomyrdin N V et al 2020 *Journal of Optics* **22** 013001.
[4] Il'ina I V, Sitnikov D S and Agranat M B 2018 *High Temperature* **56**(5) 789.

Journal of Physics: Conference Series **2067** (2021) 012015 doi:10.1088/1742-6596/2067/1/012015

[5] Fedorov V I, Serdyukov D S, Cherkasova O P et al 2017 *Journal of Optical Technology* **84**(8) 509.
[6] Fedorov V I 2017 *Biophysics* **62**(2) 324.
[7] Mattsson M-O, Zeni O and Simko M 2018 *Journal of Infrared, Millimeter, and Terahertz Waves* **39**(9) 863.
[8] Nemova E F, Cherkasova O P, Nikolaev N A and Dultseva G G 2020 *Biophysics* **65** 410
[9] Dultseva G G, Skubnevskaya G I, Tikhonov A Ya et al 1996 *J. Phys. Chem.* **100**(44) 17523.
[10] Antsygin V, Mamrashev A, Nikolaev N and Potaturkin O. 2001. *Optoelectronics, Instrumentation and Data Processing* **46** (3) 294.
[11] Williams R, Schofield A, Holder G et al 2013. *Phys. Med. Biol.* **58** 373.

Coexistence of light-induced thermocapillary and orientational effects in thin nematic films with a free surface

S.A. Shvetsov[1,2], A.S. Zolot'ko[2], G.A. Voronin[2], A.V. Emelyanenko[1], P.A. Statsenko[3], and S.I. Trashkeev[3,4]

[1] Lomonosov Moscow State University, Moscow 119991, Russia
[2] Lebedev Physical Institute, RAS, Moscow 119991, Russia
[3] Institute of Laser Physics, SB RAS, Novosibirsk 630090, Russia
[4] Voevodsky Institute of Chemical Kinetics and Combustion, SB RAS, Novosibirsk 630090, Russia

E-mail: shvetsov@polly.phys.msu.ru

Abstract. The effect of nonlinear light action on a thin ($\sim$10 μm) films of the nematic liquid crystal deposited onto the absorbing substrate is experimentally investigated. The dynamics of the orientational and thermocapillary effects is directly studied. The two types of orientational processes were found out. The first one appears for several hundreds of milliseconds when the light beam irradiation is turned on or off. The second one develops much slowly and does not relax during the light beam irradiation.

1. Introduction

The light beam absorption in isotropic liquid causes the nonlinear response due to variation of its properties [1–3]. For instance, the light beam heating can change the surface profile forming a dimple, which works as a lens for transmitted light [4, 5].This effect is mainly caused by the decrease in surface tension and the appearance of thermocapillary forces inducing the hydrodynamic flows [6–8].

Thermo-optical effects are much less investigated for the anisotropic liquids, i.e. for liquid crystals (LCs). In contrast to the ordinary liquids, LCs have partial orientational and, in some cases, translational ordering. On the macroscopic level, the average direction of the long molecular axes is defined by a unit vector, the LC director. The director field is very sensitive to various external stimuli, like electric, light and thermal fields. Thus, one can expect the interplay between the heat flux and the LC orientation when the light beam is absorbed in the medium. The most known example of thermal LC orientation is the Lehmann effect [9, 10], the director rotation in chiral nematic LCs, which, in particular, can be induced by light [11]. The light beam absorption can cause thehydrodynamic flows resulting in the director reorientation in hybrid and twist aligned nematic films [12–15]. It is interesting to note that the director deformation can be a consequence of the temperature gradient only [16, 17]. In this case, the temperature gradient field results in LC director deformation similarly to electric and magnetic fields.

In the previous study, we described the orientational effect induced by light beam heating of nematic LCs with free surfaces [18]. The nematic film was deposited onto the indium-tin-oxide (ITO) glass substrate which provides the homeotropic anchoring and particular light absorption. An air-LC interface restricts the heat outflow and the axially symmetric heat flux reorients the LC director forming an umbilical defect. As for the isotropic liquids, the dimple formation is accompanied by the LC director deformation or appears without it.

In this study, we directly observe both the light-induced surface distortion and LC director reorientation using optical microscopy methods. In particular, we focus on the moments just after turning on and off the light beam focused onto the thin nematic films with a free surface.

2. Experimental

The E7 nematic liquid crystal(Synthon Chemicals) was deposited onto the ITO-coated glass substrate. The LC droplet of the size of about 2 mm and variable thickness was held on the substrate by capillary forces. Because of the wettability of ITO-coated surface by the LC, the edge of the LC droplet can be considered as a wedge film (Fig. 1a) whose surface can be modified by a focused light beam (Fig. 1b). Both LC surfaces (LC-ITO and LC-air) provide homeotropic boundary conditions and, hence, specify the uniform orientational distribution.

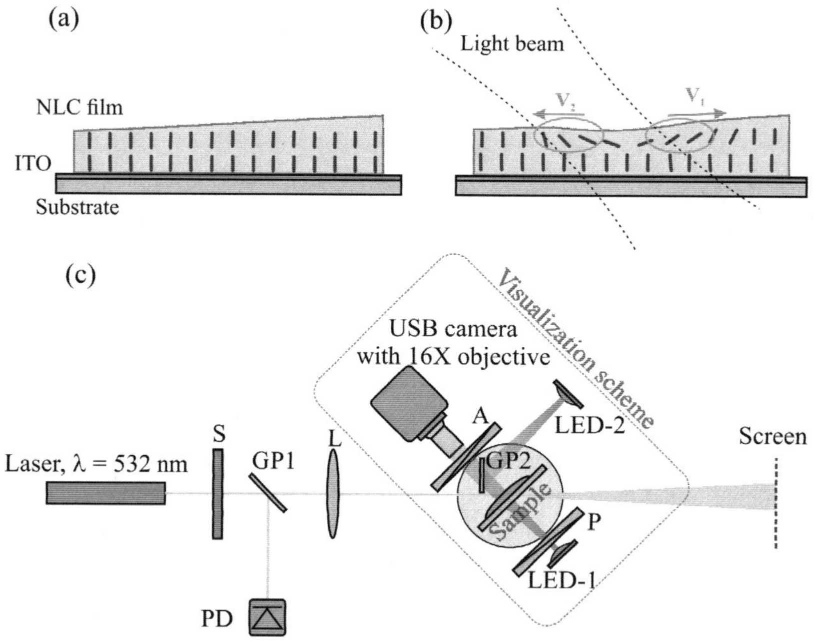

Figure 1. Schematic representation of the LC sample before (a) and during light beam irradiation (b); velocities $\mathbf{V}_1$ and $\mathbf{V}_2$ show the mass transfer from the heating area. The blue dashes show the local director orientation.The scheme of experimental setup (c): S is the shutter, L is the lens, GP1 and GP2 are glass plates, P and A polarizer and analyzer, LED-1 and LED-2 are light emitting diodes.

The optical setup (Fig. 1c) consisted of solid-state continuous-wave laser (SSP-LN-532-FN-300-0.5-LED, CNI, China) with the wavelength of $\lambda = 532$ nm, home-made mechanical shutter with switching on and off times ~ 1 ms, focusing lens, the sample with visualization scheme, and semitransparent screen. The radius of laser beam $w_0 = 26$ μm (at the intensity level of e^{-2}) was measured by the "knife-edge" method [19]. The sample was placed vertically in the beam wais.

The angle of incidence of the p-polarized light beam was 45°. The diffracted light beam was observed on the semitransparent screen placed behind the sample. The image of the illuminated area was observed with the help of a high-speed camera equipped with the 16X optical objective with a large working distance. For background illumination, two light emitting diodes (LEDs) with luminescence at $\lambda_{LED} \sim 620$ nm were used separately or simultaneously. The first LED source allows one to visualize the orientational deformation when the sample is placed between crossed polarizers (polarizer P and analizer A). The light from the second LED was partially reflected by the glass plate (GP2) and then gave the reflection image. The opened and closed shutter positions and camera recording were triggered by microcontroller. The moments of the light beam switching were controlled using a photodiode (PD) by the registration of light reflected by the glass plate (GP1) placed after the shutter. The experiments were carried out at room temperature.

3. Results and Discussion

The interference of light reflected by the LC-substrate and LC-air interfaces results in the formation of bright and dark lines (Fig. 2a-2d).Note that these interference patterns are similar to those observed for isotropic liquids [4, 5]. The LC thickness varies between bright and dark lines by the value of $\delta h = \lambda_{LED}/4n_o$, where $n_o = 1.52$ is the refractive index of an ordinary light wave. Thus, the surface profile can be reconstructed by the positions of brightness extrema along the x axis (Fig. 2e) with an accuracy of $\delta h \sim 0.1$ μm. Typical LC film profiles before and during the laser irradiation are shown in Fig. 2f. The variation of LC film thickness $|\delta h|$ with time (Fig. 2g) is measured by alternating brightness in the area of the light beam axis. The obtained data clearly show that the light beam irradiation results in the gradual variation of surface profile. The initial region in temporal dependence can be approximated by the exponential curve $|\delta h| = |\delta h_{max}|(1 - exp(-t/\tau))$ with the characteristic time $\tau \sim 3$ ms, where $|\delta h_{max}|$ is the maximum thickness change. Thus, despite of the relatively slow process of thermocapillary effect [18], the initial development is at least one order of magnitude faster.

At the LC thickness of about $h \sim 10$ μm and light beam power $P = 80$ mW, the light-induced director deformation is visualized by the bright areas in crossed polarizers (Fig. 3). Initially, only small bright regions due to the surface inhomogeneity are visible (Fig. 3a). During the light beam exposure, the bright cross appears, then becomes more intense and finally vanishes (Fig. 3b-3f). This bright cross corresponds to the axially symmetric orientational deformation (Fig.1b). Surprisingly, the bright cross is temporary observed after turning off the light beam (Fig. 3g-3l). The described effect is non-local; the size of the director deformation is several times larger than the diameter of the light beam. The appearance of the bright cross in the visualization scheme is accompanied by broadening the light beam passed through the sample.

A more complex process occurs at higher light beam power (Fig. 4). The large bright cross developes and starts to relax during several tens of milliseconds after turning on the light beam (Figs. 4a- 4e). Then the smaller bright cross manifests itself (Figs. 4f). It becomes more intense for several seconds (Figs. 4f- 4h) and does not change during the further light irradiation. The size of this cross pattern, which corresponds to the axially symmetric director deformation, is comparable to the beam diameter $2w_0$. After turning the light beam off, this deformation relaxes and, as in the previous case, the large bright cross pattern is temporary formed (Figs. 4i- 4n).

Note that the described effects do not depend on the light beam polarization and only slightly depend on the angle of incidence. The orientational deformation is symmetrical with respect to the axis which is normal to the plane of substrate in spite of the oblique incidence of the light beam.

Thus, we have two types of thermo-optical orientation in the LC film. The first one is non-steady-state and depends on the rate of the film thickness variation (Fig. 5). The increase in the light beam power leads to the more rapid variation of the film thickness as well as the

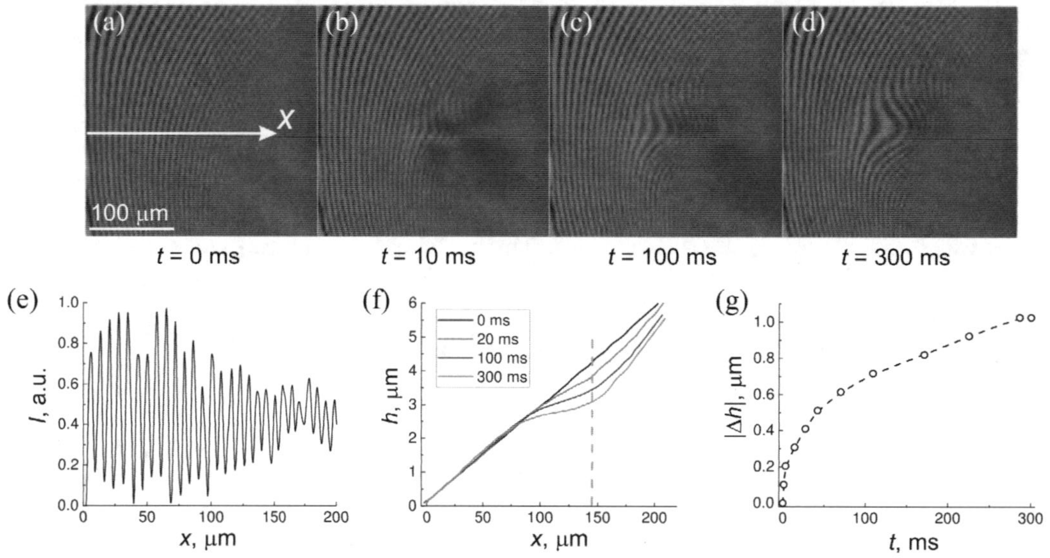

Figure 2. The images of the LC film edge (in reflected light) at different times of light beam exposure (a-d). The light beam power is $P = 120$ mW. The image brightness distribution $I(x)$ along the x axis at $t = 0$ ms (e). The reconstructed surface profiles along the x axis (the dashed line corresponds to the position of light beam axis). The time dependence of relative thickness variation $|\delta h|$ at light beam axis (g).

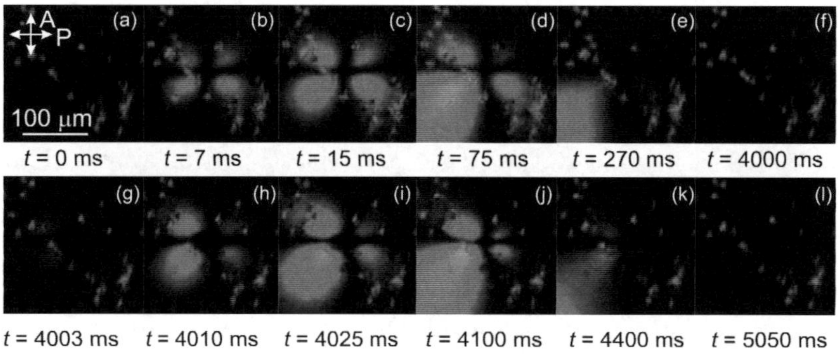

Figure 3. Polarizing microscopy images of the irradiated region (formed by the transmitted light) at different moments during and after the light beam exposure. The light beam power is $P = 80$ mW, the exposure time is $t_{exp} = 4000$ ms, the LC film thickness on the light beam axis is $h = 9.1$ μm.

growth of its absolute saturation value. We suppose, that this director deformation is due to the director alignment along the hydrodynamic flows [20]. This effect is caused by the anisotropy of translational viscosity coefficients and can be easily observed, for example, when the LC is filled into the glass cell under the action of capillary forces. The lack of positional anchoring on the free surface facilitate the hydrodynamical flow formation [21]. A slight distortion of the cross texture (see, for example, Fig. 5d) in the direction of the thickness increase can be caused by higher values of the velocity near the surface (Fig. 1b). Switching the light beam off results in the reverse flow and the director is locally oriented along it.

The second type requires sufficient light beam intensities, develops much slowly and remains

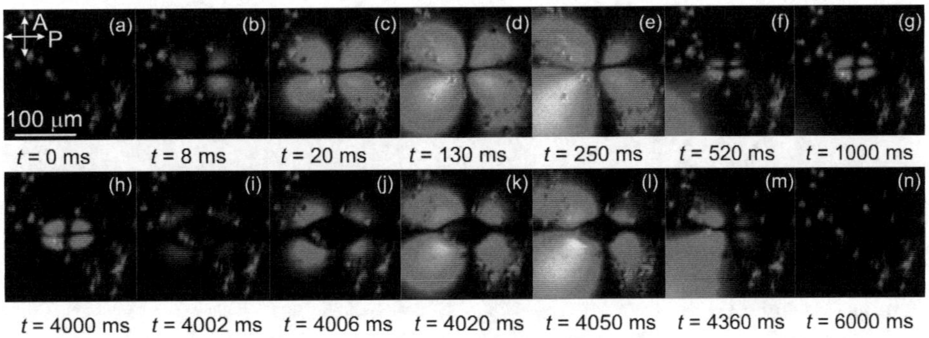

Figure 4. Polarizing microscopy images of the irradiated region (formed by the transmitted light) at different moments during and after the light beam exposure. The light beam power is $P = 120$ mW, the exposure time is $t_{exp} = 4000$ ms, the LC film thickness on the light beam axis is $h = 9.1$ μm.

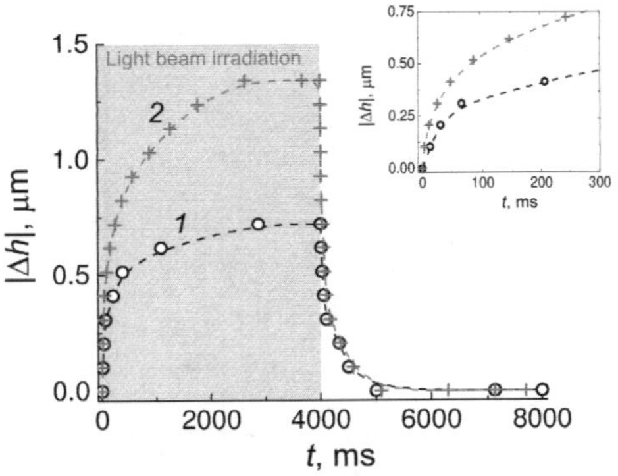

Figure 5. The time dependences of thickness variation $|\delta h|$ during and after the light beam irradiation. The light beam power is $P = 80$ (1) and 120 mW (2), the exposure time is $t_{exp} =$ 4000 ms. The inset illustrates the initial stage of thickness variation.

the same (Fig.4h) when the LC film thickness variation reaches the plateau (see curve 2 in Fig.5). This can be caused either by the temperature gradient [16, 17] or by the circular flow convection observed in LCs and polymers [22, 23].

Both types of director deformation require further theoretical study as well as computer simulations to determine the physical origin of the orientational processes more precisely. The situation can be more complex because of the interference of a number of effects, which cause the director reorientation [10].

Conclusion

To summarize, the thermo-optical orientational processes are studied experimentally for thin nematic films with free surfaces. The light absorption by the ITO layer causes both thermocapillary and orientational effects. The non-steady-state orientational deformation is observed during several hundreds of milliseconds after turning on or off the light beam. The rapid variation of the film thickness in the irradiated area indicates the development of hydrodynamic flows near the LC surface which can be responsible for this deformation. At sufficient light beam intensities, the stable LC orientation is formed. The latter one can be caused by both hydrodynamic flows and thermo-orientational effect. In each case the orientation is axially symmetric with respect to the axis oriented normally to the substrate surface.

Acknowledgments

The study was funded by the Russian Science Foundation, Grant No. 20-72-00178.

References

[1] Costa G D and Calatroni J 1978 *Applied Optics* **17** 2381–2385
[2] Da Costa G 1986 *Optical Engineering* **25** 1058–1063
[3] Da Costa G and Escalona R 1990 *Applied Optics* **29** 1023–1033
[4] Duarte-Quiroga R A and Calixto S 2000 *Appl. Opt.* **39** 3948–3954
[5] Bezuglyi B A and Fedorets A A 2001 *Technical Physics Letters* **27** 359–361
[6] Bayazitoglu Y and Lam T T 1987 *Journal of Heat Transfer* **109** 717–721
[7] Longtin J P, Jikata K H and Ogawa K 1999 *International Journal of Heat and Mass Transfer* **42** 85–93
[8] Zykov A Y and Ivanova N A 2017 *Applied Physics B* **123** 235
[9] Oswald P, Dequidt A and Poy G 2019 *Liquid Crystals Reviews* **7** 142–166
[10] Oswald P, Dequidt A and Poy G 2021 *Liquid Crystals* (Wiley) pp 117–191
[11] Bono S, Sato S and Tabe Y 2017 *Soft Matter* **13** 6569–6575
[12] Akopyan R S, Alaverdian R B, Santrosian E A and Chilingarian Y S 2001 *Journal of Applied Physics* **90** 3371–3376
[13] Hakobyan R S, Alaverdyan R B, Gevorgyan G S and Kirakosyan A A 2010 *JETP Letters* **92** 457–459
[14] Aleksanyan A, Gevorgyan G, Hakobyan R and Alaverdyan R 2012 *International Journal of Modern Physics: Conference Series* **15** 120–128
[15] Hakobyan M R, Alaverdyan R B and Hakobyan R S 2014 *Molecular Crystals and Liquid Crystals* **596** 152–162
[16] Demenev E I, Pozdnyakov G A and Trashkeev S I 2009 *Technical Physics Letters* **35** 674–677
[17] Trashkeev S I and Britvin A V 2011 *Technical Physics* **56** 747–753
[18] Shvetsov S A, Zolot'ko A S, Voronin G A, Emelyanenko A V, Avdeev M M, Bugakov M A, Statsenko P A and Trashkeev S I 2021 *Optical Materials Express* **11** 1705
[19] Khosrofian J M and Garetz B A 1983 *Applied Optics* **22** 3406
[20] Belyaev V V 2001 *Physics-Uspekhi* **44** 255–284
[21] Pan R P, Hsiung H and Shen Y R 1987 *Phys. Rev. A* **36** 5505–5508
[22] Dradrach K, Bartkiewicz S and Miniewicz A 2016 *Physical Chemistry Chemical Physics* **18** 3832–3837
[23] Choi H and Takezoe H 2016 *Soft Matter* **12** 481–485

Semiconductor quantum well based shutters for NIR laser mode-locking with ~ GHz repetition rate

N N Rubtsova[1], A A Kovalyov[1], D V Ledovskikh[1], V V Preobrazhenskii[1], M A Putyato[1], B R Semyagin[1], S A Kuznetsov[2], V S Pivtsov[2]

[1]Institute of Semiconductor Physic, Siberian Branch of Russian Academy of Sciences, Novosibirsk, 13, Acad. Lavrentyev av., Russian Federation
[2]Institute of Laser Physics, Siberian Branch of Russian Academy of Sciences, Novosibirsk, 15B, Acad. Lavrentyev av., Russian Federation

E-mail: rubtsova@isp.nsc.ru

Abstract. Fast semiconductor shutters based on coupled wells were designed in the search for reliable, compact and cheap key element of GHz repetition rate NIR lasers passive mode-locking. Stable 0.98 GHz repetition rate 200-fs Yb:KYW laser pulses were demonstrated for SESAM including semiconductor reflector and a layer of quantum wells. The damage threshold estimate for the SESAM is ~ 8.87 mJ/cm^2. Other type of shutter – DSAM – was developed with dielectric reflector and the layer of quantum wells transferred over reflector. The measured recovery time was about 2-3 ps for both types of saturable absorbers. The efficiency relative to the incident pump power was 57% for the SESAM and 19% for the DSAM. Average output power of 2.54 W for the all-semiconductor shutter (SESAM) and of 0.92 W for the dielectric mirror with a saturable absorber (DSAM) were obtained. Actual state of the art for the shutters design is considered.

1. Introduction

Femtosecond lasers are requested in many areas of fundamental science and applications. Their applications field covers research of fast transient processes in various media and in a wide spectral area, in biological and medical imaging (in particular, optical coherent tomography), in optical communication, in high-precision coordinate-temporal measurements of ground and cosmic location objects, in metrological support of high-technology branches of industry, etc. Briefly speaking, compact, reliable, cost-effective ultra-short pulse lasers are requested.

An idea of how to get ultra-short pulse is known since 70-th of the past century [1] – the synchronization of longitudinal modes of laser (mode locking technique); this method was firstly demonstrated in the system “dye laser + saturable absorber solution”. Today we know several approaches to the mode locking. Active mode locking can be realized by placing acousto-optic modulator inside a laser cavity, and this technique is well suited to high gain lasers. Passive mode locking can use Kerr lens effect; this approach is sensitive to laser cavity adjustment, and it needs high intensity of radiation to create remarkable beam deformation because of nonlinear change of refraction index. Passive mode locking via saturable absorption acquired new life due to idea to apply SEmiconductor Saturable Absorption Mirror (SESAM) [2]. Today, the SESAM became a key element of laser mode locking, and it provides self-starting self-sustainable mode locking regime without Q-switch for a wide class of lasers from NIR to MIR.

To introduce our activity in this area of research, some examples of SESAMs designed for lasers with low repetition rate of ultra-short pulses will be given below. From the large variety of semiconductor nanostructures (quantum wells, quantum dots, dots-in-wells etc.), we chose the quantum wells (QW). For application in compact femtosecond lasers, the QW based SA are the most reliable, because they can operate at much higher working radiation intensity.

2. Shutter based on single quantum well

At the first step, we chose a QW of the 1-st type (both electrons and holes are confined) with ideal lattice matching between the well and the barrier materials. This is a quite rear case. Usually, there is no lattice matching, and the quantum well layers are under mechanical stress. Fig.1 shows energy diagram of single quantum well InGaAs between InP barriers. The band gaps are indicated for materials of barriers (InP) and quantum well ($In_{0.53}Ga_{0.47}As$). Dashed arrow shows transition between minimal energy quantization levels of heavy hole (1hh) and electron (1e). The possibility of exciton localization in this QW seems attractive because the excitonic transition is easy saturable.

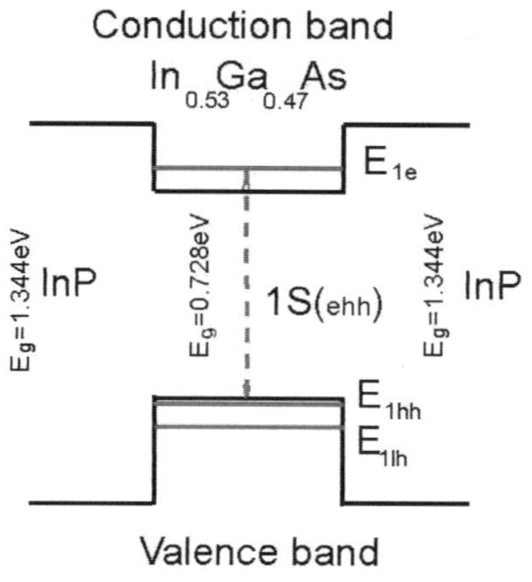

Fig.1. Energy diagram of the first type single quantum well.

SESAM mirror for Cr:YAG laser (1.4 – 1.6 micrometers) was produced via hybrid technology (1-st type single QW was transferred onto dielectric reflector and sapphire substrate, and had an ant-reflective coating from the QW side. The $In_{0.53}Ga_{0.47}As$ layer of 6 nm thickness is single QW between InP barriers. QW is located at the maximum of standing wave pattern. By using this shutter, and by thorough group delay dispersion (GDD) compensation, 27 fs pulses were obtained, which corresponded to 5 oscillations of carrier frequency only [3]. However, the output power of laser was low, and a saturable loss was comparable with non-saturable one.

3. Shutter based on several isolated quantum wells

As the next step, the combination of several functions (semiconductor total reflector + saturable absorber + GDD compensator) was realized in SESAM for Nd^{3+}:KGd$(WO_4)_2$ (Nd:KGW) laser (central wavelength of 1.06 microns) [4]. In this case, two quantum wells of $In_{0.25}Ga_{0.75}As$ were used, and they were separated by thick (of about 23 nm) barriers of GaAs. Stable mode locking regime was provided by this all-in-one all-semiconductor mirror. Repetition rate was 70 MHz, and pulse duration was of about 0.6 ps. Laser crystal Nd:KGW is not very efficient; besides, slow trend of mode locked pulses was observed, probably, due to thermal lens generated in the laser crystal.

Ytterbium lasers Yb^{3+}:KY$(WO_4)_2$ operating around 1.04 microns demonstrate high efficiency with diode laser pumping. In collaboration with colleagues from Belarus, we developed SESAM for Yb:KYW laser (look [4]) based on several isolated quantum wells separated by thick barriers which prevents overlapping of wave functions from neighbor wells. Its kinetics could be approximated by bi-exponential low with two distinct relaxation times: τ_1 corresponding to ionization of photo-generated excitons localized in quantum wells (usually about 0.4 ps), and τ_2 which reflects final relaxation due to electron-hole (*e-h*) recombination (for this SESAM, about 200 ps). Just the relaxation time τ_2 (known also as recovery time) determines either the shutter operation is enough fast. For this SESAM, laser intensity provoking optical damage was estimated as 10 GW/cm^2. Further, we used Yb:KYW laser in our work [5].

When thinking about shutters for the compact lasers, it was necessary to accelerate *e-h* recombination, recovery time of SESAM should be shorter, SESAM should be faster.

Acceleration of *e-h* recombination can be obtained by creation of additional electron-hole recombination centers (in particular, point defects). This can be done via post-growth irradiation of the shutter by heavy ions beam [6], or by doping of well and barrier materials by another atoms [7], or by intentional defects generation in a course of molecular beam epitaxy (MBE) due to lattice mismatching [8]. All above mentioned approaches induce partial degradation of the sample crystal lattice. This, in turn, reduces (or eliminates totally) contribution of localized excitons into absorption spectrum. However, just this contribution is able to provide absorption saturation at low intensity of working radiation (so called "easy start of mode locking").

We carried out experiment using another way of *e-h* recombination acceleration; we investigated UV laser post-growth treatment of SESAM. For this goal, excimer XeCl laser operating at the wavelength of 308 nm was used (UV quantum energy was 4.03 eV, pulse duration 15 ns, maximum intensity of 13 MW/cm^2, repetition rate of 1 Hz). Effect of UV radiation was detected via photoluminescence (PL) spectra and via reflectivity kinetics of the samples before and after irradiation. UV laser modification of PL and kinetics change proved threshold-like process [9]. Recovery time shortened from 100-200 ps to 10-20 ps (almost by two orders of magnitude). We carried out X-ray diffraction testing; X-ray source operated at the wavelength of 0.154 nm with the beam size 1.5mm×0.1 mm. Monotonous changes in diffractive curves can not be attributed to the influence of UV irradiation, that is rather initial MBE growth in-homogeneity. Hence, the structure changes take place in a thin surface layer of about 5-10 nm thickness (not detectable for X-ray diffraction). This value correlates with the penetration depth of UV laser radiation: approx 13 nm (estimated by using absorption coefficient 7.72×10^5 cm^{-1}). So, the mechanism of *e-h* recombination acceleration proved the same as in the previous works cited above, i.e. generation of point defects, in this case – assisted by UV laser [10].

The results on the post-growth UV laser treatment of SESAMs seemed promising. But, the nature of recovery time shortening is again the *e-h* recombination acceleration due to point defects generated by UV laser radiation. The semiconductor lattice is damaged; besides, we don't know the behavior of point defects under high intensity working NIR radiation.

We needed some other approach to recovery time shortening which maintains ideal crystal quality of semiconductor structure. The first step was design of quantum wells separated by super-lattice barriers.

4. Shutter based on quantum wells with super-lattice barriers

Low band gap material $In_{0.25}Ga_{0.75}As$ insertions into the barriers should promote the charge carriers tunneling between the quantum wells, and this should accelerate the recovery of optical shutter. Absorption spectra of layer of quantum wells $In_{0.25}Ga_{0.75}As$ (15 nm thickness) separated by thick barriers AlAs (23 nm thickness) showed narrow spectral peak (about 20 nm spectral width) attributed to excitonic contribution. The layer of the same quantum wells separated by super-lattice barriers (quantum well 15 nm of $In_{0.25}Ga_{0.75}As$ and barrier consisted of 3ML AlAs/8ML well/2ML AlAs/8ML well/2MLAlAs/8ML well/3MLAlAs, where ML means monolayer, and well is $In_{0.25}Ga_{0.75}As$) had much wider spectrum (of 100 nm spectral width), and its kinetic gave the recovery time of about 8 ps [11]. On the base of this nanostructure, SESAM for Yb:KYW laser with super-lattice barriers and low non-saturable loss was made [12]. Pulses were close to spectral-limited ones with the time-bandwidth product of about 0.319. Maximum average power was 1.57 W, which is comparable to CW regime power of 1.62 W. Pulse repetition rate was 70 MHz with the peak power of 171 kW and pulses duration of 115 fs.

Journal of Physics: Conference Series **2067** (2021) 012017 doi:10.1088/1742-6596/2067/1/012017

5. Shutters for compact NIR lasers

Taking into account previous experience, we can generate a list of requirements to optical shutters for passive mode locking in compact lasers. The shutter should provide saturable absorption (modulation depth) of necessary level; it should reveal fast saturation and recovery (especially for femtosecond lasers with fast repetition rate); it should demonstrate high reflectivity in spectral area of the laser gain (possess wide spectral range and high quality reflectors); it should be of low non-saturable loss (be of good optical quality); it should be resistive to intense working radiation; it would be desirable to obtain shutter of long lifetime and of low cost, etc.

Actually, two basic constructions of shutters for compact Yb:KYW laser were created. The first one is monolithic SESAM including semiconductor reflector grown over GaAs (001) epi-ready substrate (made of 20 pair of quarter-wavelength layers of GaAs and AlAs) and a layer of 3 quantum wells (27 ML of $In_{0.25}Ga_{0.75}As$) separated by thin barriers (5 ML of GaAs). This shutter could provide maximum modulation depth of 10%, which was tested by linear absorption spectrum of the sample covered by anti-reflecting layer of Y_2O_3 (of quarter-wavelength thickness). In order to adopt the shutter to a given Yb:KYW laser, upper dielectric mirror was sputtered over the anti-reflecting coating (in our case, we used alternation of ZrO_2 and SiO_2 quarter-wavelength layers); so, when finished by additional dielectric layers, the structure gives modulation depth of about 1%. Another shutter, DSAM, included dielectric reflector attached to metal substrate, and the same layer of 3 quantum wells transferred onto dielectric reflector.

Each of the shutters, SESAM and DSAM, has some advantages and drawbacks. Let's consider specific features of reflectors used in the two shutters. For SESAM with 3 QW as saturable absorber, reflectivity optical table is of 110 nm width. DSAM shows reflectivity optical table is of 210 nm width. SESAM manifests lateral in-homogeneity of reflectivity optical table spectral position. Dielectric reflectors prepared via sputtering in vacuum show high lateral homogeneity. So, dielectric mirror has reflectivity table width twice larger than the semiconductor one, it is homogeneous, and semiconductor mirror (MBE grown of super-clean materials) is much more expensive.

To estimate how fast the shutters are, a single wavelength technique pump-probe with double modulation was used [13]. Saturating and probe beams of Yb:KYW laser (FL-1000, Minsk, Belarus, 1040 nm central wavelength, 130 fs pulse duration, repetition rate about 70 MHz) were chopped by mechanical choppers at two different non-multiple frequencies F_1 and F_2 stabilized at the accuracy of 1 Hz. The probe radiation pulse could be delayed relative the saturating pulse by time interval from -10 ps to +40 ps. The idea to detect nonlinear response of a sample reflectivity at the sum frequency F_1+F_2 was realized via generation of the reference signal at the sum frequency (from reference signals generated by optron pairs of each chopper) and by application of radio-filter at the sum frequency to detect the probe beam radiation. This approach permitted to eliminate completely scattered radiation of saturating beam. The set-up allowed us to get the minimal relative reflectivity change detection of about 5×10^{-6}.

Kinetic curves of quantum wells reflectivity show bi-exponential decay in the limit of low intense saturating radiation only. A phenomenological model of the kinetics of reflection of a semiconductor mirror with quantum wells was developed, which takes into account the contribution of two subsystems: excitons and electron-hole pairs [14].

The model makes it possible to compare with experimental data. For SESAM and DSAM designed for compact Yb:KYW laser we got the following relaxation parameters: τ_1=0.4 ps (ionization time of excitons), τ_2= 2-3 ps (electron-hole recombination time), and τ_3=0.8 ps (quasi-equilibrium installation time in manifold of un-correlated electrons and holes) [15].

Both types of shutters, SESAM and DSAM, were tested in compact Yb:KYW laser operating under optical pumping of CW diode laser at 980 nm. GDD was compensated by two elements based on Gires-Tournois interferometers. For both shutters, stable mode locking regime was obtained, and pulses duration was about 200 fs [15].

Efficiency of both shutters was estimated at 720 MHz repetition rate. For our SESAM, efficiency was 57% (some higher than for commercial SESAM), and maximum average output power was of 2.54 W. DSAM proved some worse: 19% of efficiency, and maximum average output power was of 0.92 W.

For SESAM, the shortening of laser cavity was carried out up to repetition rate of 970 MHz, and mode locking regime retained stable with high signal-to-noise ratio for central spectral component corresponding to the inter-mode distance. Side-band patterns appeared near the central component at repetition rate of 982.65 MHz. Parameters of working radiation under these conditions were used to estimate the optical damage threshold for our SESAM at the level of surface energy density of about 8.9 mJ/cm^2.

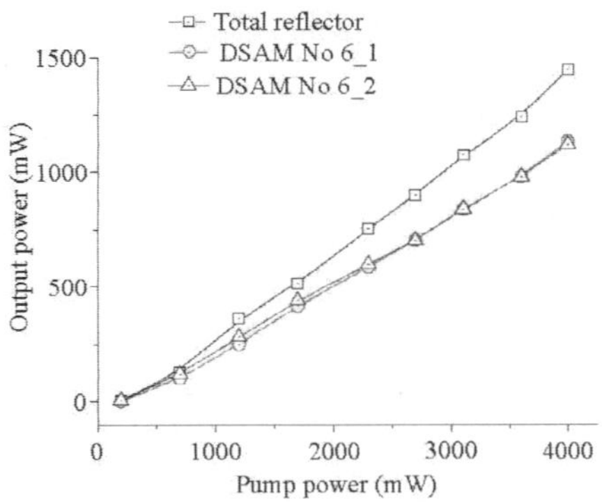

Fig.2. Comparison of average output power in mode locking regime with 2 samples of DSAM with output power in CW regime with total reflector.

For DSAM, the optical quality can be estimated by comparison of the average output power in mode locking regime with output power in CW regime with total reflector placed in spite of DSAM. Results presented in Fig. 2 demonstrate good reproducibility of two DSAM samples. However, they are still far from the limit of low non-saturable loss.

6. Conclusion

To summarize, we can state the following. Stable mode-locking up to 0.98 GHz repetition rate in Yb:KYW laser was demonstrated with SESAM based on coupled quantum wells (with thin barriers). Layer of quantum wells separated by thin barriers transferred onto dielectric reflector showed recovery time acceptable for passive mode locking in compact femtosecond lasers of NIR spectral area. DSAM needs further improvement, but it is promising in design of cheap and fast optical shutters for NIR lasers mode-locking.

Acknowledgments

The study was carried out with the financial support of the Russian Foundation for Basic Research grant No. 18-29-20007. S A Kuznetsov and V S Pivtsov acknowledge partial support from the RFBR Grant № 19-02-00242, the Ministry of Science and Higher Education of the Russian Federation (№ 121033100064-9), and the Center for Collective Use “Laser Femtosecond Complex” for hardware support.

References

[1] *Ultrashort Light pulses* 1977 Ed. S L Shapiro (Springer)
[2] Keller U, Weingarten K J, Kärtner F X, Kopf D, Braun B, Jung I D, Fluck R, Hönninger C, Matuschek N, Aus der Au J. 1996 *IEEE J. Selected Topics in Quantum Electronics (JSTQE)* **2** 435

[3] Rubtsova N N, Preobrazhenskii V V, Kovalyov A A, Putyato M A, Pchelyakov O P, Sorokin E, Naumov S, Sorokina I T 2003 A3B5-dielectric nanostructures for near and mid infrared short pulse solid state lasers *12-th International Symposium "Nanostructures: Physics and Technology", St. Petersburg, 21-25 June 2003,* p 297

[4] Rubtsova N N, Kuleshov N V, Kisel' V E, Kovalyov A A, Kurilchik S V, Preobrazhenski V V, Putyato M A, Pchelyakov O P, and Shamirzaev T S 2004 *Laser Physics* **19** 285

[5] Kuznetsov S A and Pivtsov V S 2014 *Quantum Electronics* **44** 444

[6] Mangeney J, Oudar J L, Harmand J C, Meriafec C, Patriarche G, Aubin G, Stelmakh N, Lourtioz J M 2000 *Appl. Phys. Lett.* **76** 1371

[7] Tsuyoshi Okuno, Yasuaki Masumoto 2001 *Appl. Phys. Lett.* **79** 764

[8] Soumalainen S, Vainionpaa A, Tengvall O, Hakulinen T, Karirinne S, Guina M, Okhotnikov O G 2005 *Appl. Phys. Lett.* **87** 121106

[9] Rubtsova N N, Kuleshov N V, Kisel' V E, Kochubei S A, Kovalyov A A, Kurilchik S V, Preobrazhenskii V V, Putyato M A, Pchelyakov O P, Shamirzaev T S 2010 *Laser Physics* **20** 1262

[10] Kashkarov P K, Timoshenko V Yu 1995 *Poverkhnost'. Fizika, Khimiya, Mekhanika* **6** 1 (in Russian)

[11] Kovalyov A A, Kuleshov N V, Kisel' V E, Kuril'chik S V, Preobrazhenskii V V, Putyato M A, Pchelyakov O P, Rubtsova N N, Semyagin B R 2016 *Laser Physics* **26** 025001

[12] Kovalyov A A, Preobrazhenskii V V, Putyato M A, Pchelyakov O P, Rubtsova N N, Semyagin B R, Kisel' V E, Kuril'chik S V, Kuleshov N V 2011 *Las. Phys. Lett.* **8** 431

[13] Gol'dort V G, Kovalyov A A, Ledovskikh D V, Rubtsova N N 2018 *Instruments and Experimental Techniques* **61** 94

[14] Borisov G M, Gol'dort V G, Kovalyov A A, Ledovskikh D V, Rubtsova N N 2017 *Sibirskii Fizicheskii Zhurnal* **12** 109 (In Russian)

[15] Rubtsova N N, Kovalyov A A, Ledovskikh D V, Preobrazhenskii V V, Putyato M A, Semyagin B R, Kuznetsov S A, Pivtsov V S, and Semenko A V 2020 *Laser Physics* **30** 025001

Journal of Physics: Conference Series **2067** (2021) 012018 doi:10.1088/1742-6596/2067/1/012018

Theoretical bases of combined nonlinear optical demodulator-amplifier for broadband microwave photonics system

G N Nikolaev, S L Mikerin

Institute of Automation and Electrometry, Siberian Branch of Russian Academy of Science, Academician Koptyug ave. 1, Novosibirsk 630090, Russia

E-mail: nikolaev@iae.nsk.su

Abstract. Our demodulator is based on the use of a strong light pump wave along with a signal modulated light wave. The demodulator is linear in terms of the modulation signal, which makes it possible to obtain a demodulated signal without harmonic distortion. The modulator conversion efficiency and signal gain increase with increasing microwave baseband.

1. Introduction

Microwave photonics is based on the use of laser radiation, modulated by a microwave signal, in the tasks of receiving, transmitting and processing information [1, 2]. The transition to microwave photonics provides noise immunity, information security, broadband, reduced noise distortions during signal processing, increased frequency selectivity, compact hardware implementation. This allows the creation of devices with parameters unattainable for traditional microwave electronics. For ultrafast microwave photonic systems, high-speed light modulators and demodulators are critical. The efficiency of such existing devices drops sharply after exceeding their critical microwave frequency, which for the fastest of them is about 100 GHz [3, 4, 5].

We have proposed a nonlinear optical demodulator, the efficiency of which increases with an increase in the signal modulation frequency.

2. Proposed Nonlinear Optical Demodulation Scheme

2.1. Desired properties of a nonlinear optical demodulator

We would the nonlinear optical demodulator have the following properties:

- it would be linear to the information signal,
- the hardware components of the modulation path could be used in its design,
- it would also be an amplifier at the same time.

2.2. Technological implementation: nonlinear optical planar waveguide inside the strip.

To implement such an optical demodulator, we propose to use the same design as for a nonlinear optical phase light modulator: a plane nonlinear optical waveguide inside a stripline.

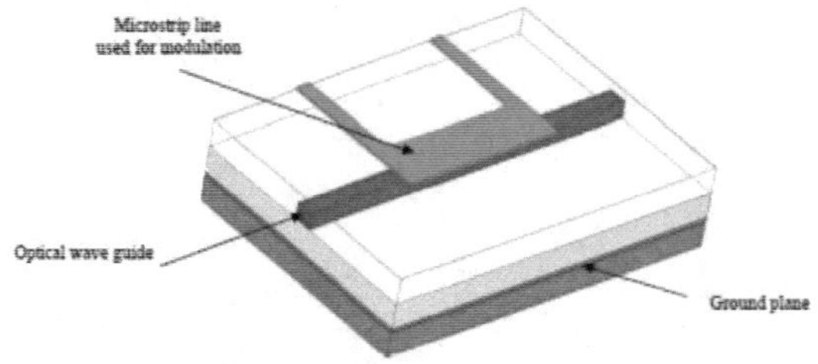

Figure 1: Planar nonlinear optical waveguide with microstrip line.

Structurally, such a demodulator is a planar nonlinear single-mode optical waveguide inside a microstrip line. (see Figure 1).

Both a modulated signal light wave and a much more intense light pump wave, which frequency coincides with the carrier frequency of the signal wave, are fed to the input of the waveguide in the direction of axis z. A microwave wave arises as a result of the nonlinear optical interaction of these waves. It propagates along the stripline in the same direction. Because the intensity of the pump wave is much higher than the intensity of the signal wave (so the amplitude of the pump wave can be considered constant), the amplitude of the microwave wave turns out to be linearly related to the amplitude of the modulated light wave. Therefore there are no nonlinear distortions in such a demodulation scheme.

3. Theoretical description of nonlinear demodulator

3.1. Equations for dimensionless wave amplitudes

Due to the smallness of nonlinearity of the planar optical waveguide, wave amplitudes vary slightly at distances of the order of their wavelengths. Therefore, the process of nonlinear optical demodulation may describe in the slowly varied amplitudes approximation.

We assume that all waves have linear polarization along the x-axis perpendicular to the planar waveguide. In this case, the nonlinear optical conversion in the optical waveguide core depends on the largest component $\chi^{(2)}_{xxx} \equiv \chi^{(2)}$of the second-order nonlinear susceptibility tensor. For the case of harmonic amplitude modulation of a signal wave, the following differential couple-waves equations describe the demodulation process:

$$\frac{\partial \bar{a}}{\partial \tau} + iq\bar{a} = i\left[C_+ a^* a_+ + C_- a\, a_-^*\right], \tag{1}$$

$$\frac{\partial a_+}{\partial \tau} = ic_+ a\bar{a}, \tag{2}$$

$$\frac{\partial a_-^*}{\partial \tau} = -ic_-^* a^* \bar{a}, \tag{3}$$

where a, a_- (a_+), and $\overline{a}$ are dimensionless amplitudes of the pump optical wave, the Stokes (anti-Stokes) spectral component of the modulated optical wave, and the microwave wave, respectively;

$$C_\pm = \sqrt{\frac{(\omega \pm \Omega)}{\omega} \frac{n(\omega)}{n(\omega \pm \Omega)} \frac{\chi^{(2)}(\Omega; \mp\omega, \Omega \pm \omega)}{\left|\chi^{(2)}(0; \omega, -\omega)\right|}}, \tag{4}$$

$$c_\pm = \sqrt{\frac{(\omega \pm \Omega)}{\omega} \frac{n(\omega)}{n(\omega \pm \Omega)} \frac{\chi^{(2)}(\omega \pm \Omega; \omega, \pm\Omega)}{\left|\chi^{(2)}(0; \omega, -\omega)\right|}}, \tag{5}$$

$$q = \left(\frac{V_p - v_g}{v_g}\right)\varkappa L_{NL}, \tag{6}$$

$$L_{NL}^{-1} \equiv 2\pi|\chi^{(2)}(0;\omega,-\omega)|\frac{|E|}{|a|}\sqrt{\frac{k}{n(\omega)}\frac{\varkappa}{n(\Omega)}}, \tag{7}$$

L_{NL} is non-linear interaction length, $\tau = z/L_{NL}$; ω,k and $n(\omega)$ are circular frequency, wave number in vacuum and refractive index of the pump wave correspondingly, and Ω,$\varkappa$ and $n(\Omega)$ are corresponding quantities for the microwave wave; E is the electric field amplitude of the pump wave; V_p and v_g are phase velocity of the microwave wave and group velocity of the light one, correspondingly. The squares of the moduli of the dimensionless amplitudes of the strong reference (a), modulated $(a_\pm)$ and microwave $(\bar{a})$ waves are equal to the flux fractions of the corresponding quanta from the total quanta flux $\mathcal{F}_0$:

$$|a|^2 = \frac{cn(\omega)|E|^2}{8\pi\hbar\omega}\frac{1}{\mathcal{F}_0}, \tag{8}$$

$$|a_\pm|^2 = \frac{cn(\omega\pm\Omega)|E_\pm|^2}{8\pi\hbar(\omega\pm\Omega)}\frac{1}{\mathcal{F}_0}, \tag{9}$$

$$|\bar{a}|^2 = \frac{cn(\Omega)\left|\overline{E}\right|^2}{8\pi\hbar\Omega}\frac{1}{\mathcal{F}_0}. \tag{10}$$

Moduli of dimensionless amplitudes are satisfy the follows relations:

$$|a_\pm|^2, |\bar{a}|^2 \ll |a|^2 \approx 1. \tag{11}$$

According to the first equation (1) of the system (1) – (3), the change in the amplitude of the microwave wave is determined by two processes described by two terms in square brackets on the right side of this equation. The first term describes annihilation of the anti-Stokes light quantum with the frequency of $\omega+\Omega$ and simultaneous creation of the light quantum with the frequency of ω and the microwave quantum with the frequency of Ω. The second term describes annihilation of the light quantum with the frequency of ω and the simultaneous creation of the anti-Stokes light quantum with the frequency $\omega-\Omega$ and the microwave quantum with the frequency Ω (see Figure 2).

From the system (1) – (3), we obtain a closed equation for the amplitude of the microwave wave by eliminating the amplitudes of the Stokes and anti-Stokes optical waves from the first equation (1) of the system by means of the the second (2) and third (3) equations:

$$\frac{\partial^2\bar{a}(\tau)}{\partial\tau^2} + iq\frac{\partial\bar{a}(\tau)}{\partial\tau} + g\bar{a}(\tau) = 0, \tag{12}$$

where

$$g \equiv (C_+c_+ - C_-c_-)|a|^2. \tag{13}$$

The boundary conditions for the equation (12) are

$$\bar{a}(0) = 0,$$

$$\frac{\partial\bar{a}(\tau)}{\partial\tau}|_{\tau=0} = i\left[C_+a^*a_+(0) + C_-a\,a_-^*(0)\right].$$

There are possible three qualitatively different operation modes of the proposed nonlinear optical demodulator. Just the sign of the combination $\left(g+\frac{q^2}{4}\right)$ defines the mode.

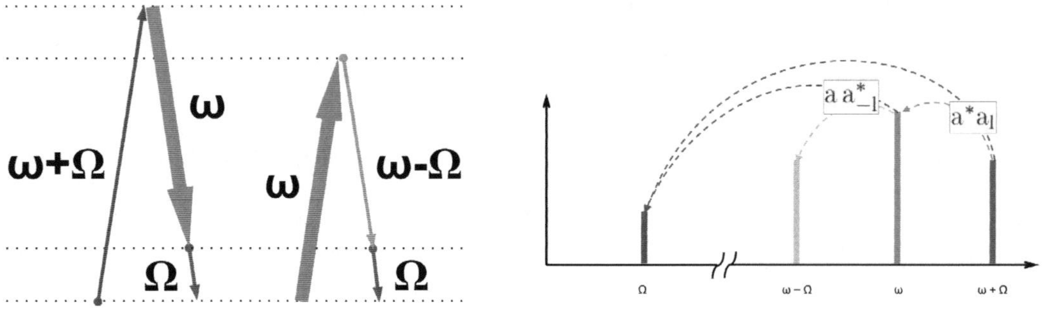

(a) (b)

Figure 2: Energy diagrams and spectral energy transfers illustrating nonlinear optical demodulation. (a) A microwave quantum is created both during the decay of an anti-Stokes photon (left diagram) and during the decay of a pump wave photon (right diagram). (b) Frequency spectrum of nonlinearly interacting waves and directions of energy transfer.

3.2. Oscillatory mode ($g + \frac{q^2}{4} > 0$)

Under condition $g + \frac{q^2}{4} > 0$, the solution to equation (12) can be represented as

$$\bar{a}(\tau) = iB \frac{\sin\left(\sqrt{\left(g + \frac{q^2}{4}\right)}\tau\right)}{\sqrt{\left(g + \frac{q^2}{4}\right)}} \exp\left(-i\frac{q}{2}\tau\right), \tag{14}$$

where

$$B \equiv \left[C_+ a^* a_+(0) + C_- a\, a_-^*(0)\right]. \tag{15}$$

It is convenient to introduce the dimensionless nonlinear coherence length τ_c:

$$\tau_c \equiv \frac{\pi}{2} \frac{1}{\sqrt{\left|g + \frac{q^2}{4}\right|}}.$$

In terms of the nonlinear coherence length τ_c, expression (14) for $\bar{a}(\tau)$ takes the form

$$\bar{a}(\tau) = i\frac{2}{\pi} B\tau_c \sin\left(\frac{\pi}{2}\frac{\tau}{\tau_c}\right) \exp\left(-i\frac{q}{2}\tau\right).$$

The microwave amplitude modulus $|\bar{a}(\tau)|$ is maximal at $\tau = \tau_c$ and is equal to $2B\tau_c/\pi$. In the case of $\tau \ll \tau_c$, we have $|\bar{a}(\tau)| \approx |B|\,\tau$.

The quantum efficiency η of the nonlinear optical modulator is the ratio of the flux of quanta of the microwave wave at the output of the modulator to the sum of quanta fluxes of the Stokes and anti-Stokes components of the modulated wave at the input of the modulator:

$$\eta = \frac{|\bar{a}(\tau)|^2}{|a_+(0)|^2 + |a_-(0)|^2} = \frac{\left|C_+ a^* a_+(0) + C_- a\, a_-^*(0)\right|^2}{|a_+(0)|^2 + |a_-(0)|^2} \frac{\sin^2\left(\sqrt{\left(g + \frac{q^2}{4}\right)}\tau\right)}{\left(g + \frac{q^2}{4}\right)}. \tag{16}$$

In the case of $\tau \ll \tau_c$, it is approximately equal to

$$\eta \approx \frac{\left|C_+ a^* a_+ (0) + C_- a\, a_-^* (0)\right|^2}{\left|a_+ (0)\right|^2 + \left|a_- (0)\right|^2} \tau^2. \tag{17}$$

3.3. Parametric gain mode ($g + \frac{q^2}{4} < 0$)

In this case, the expression for the dimensionless amplitude of the microwave wave takes the form

$$\bar{a}(\tau) = iB \frac{\sinh\left(\sqrt{\left|g + \frac{q^2}{4}\right|}\tau\right)}{\sqrt{\left|g + \frac{q^2}{4}\right|}} \exp\left(-i\frac{q}{2}\tau\right).$$

One can see that the amplitude of the microwave wave increases exponentially at large $\sqrt{\left|g + \frac{q^2}{4}\right|}\tau$ in increasing the modulator length τ.

The quantum efficiency η of the nonlinear optical modulator for this case is as follows:

$$\eta = \frac{\left|C_+ a^* a_+ (0) + C_- a\, a_-^* (0)\right|^2}{\left|a_+ (0)\right|^2 + \left|a_- (0)\right|^2} \frac{\sinh^2\left(\sqrt{\left|g + \frac{q^2}{4}\right|}\tau\right)}{\left|g + \frac{q^2}{4}\right|}.$$

It also exponentially increases at large $\sqrt{\left|g + \frac{q^2}{4}\right|}\tau$ in increasing the modulator length τ. In the opposite case, when $\sqrt{\left|g + \frac{q^2}{4}\right|}\tau \ll 1$, it takes the already known expression (17).

3.4. Linear gain mode ($g + \frac{q^2}{4} = 0$)

The mode is intermediate between the two modes discussed above. In this case, $\tau_c = \infty$, and the amplitude of the microwave wave increases linearly in increasing the length τ of the modulator:

$$\bar{a}(\tau) = iB\tau \exp\left(-i\frac{q}{2}\tau\right).$$

The quantum efficiency η, on the other hand, grows quadratically in increasing the length τ of the modulator:

$$\eta = \frac{\left|C_+ a^* a_+ (0) + C_- a\, a_-^* (0)\right|^2}{\left|a_+ (0)\right|^2 + \left|a_- (0)\right|^2} \tau^2.$$

Both regimes considered above tend to this regime in the limit $\sqrt{\left|g + \frac{q^2}{4}\right|}\tau \longrightarrow 0$.

4. Estimations and discussion

So, we have shown that the dynamics of the nonlinear optical demodulator and its mode are determined by the sign and value of g. In the absence of frequency dispersion g = 0, as follows from the definition of g (13) and expressions (4) and (5). Let us take into account the frequency dispersion in the first approximation in $\Omega/\omega \ll 1$:

$$g \approx 2\left[\left(1-\frac{V_p-v_g}{v_g}\right)+2D\right]|a|^2\frac{\Omega}{\omega}; \quad (18)$$

$$D \equiv \frac{\omega}{\left|\chi^{(2)}(0;-\omega,\omega)\right|}\frac{\partial\chi^{(2)}(-y+\omega;-\omega,y)}{\partial y}|_{y=\omega} \quad (19)$$

The derivative in (19) one can estimate as $\chi^{(2)}(0;-\omega,\omega)/\omega$. According to this estimate, $D=1$. Thus, the expression for g takes the form:

$$g \approx 2\left[3-\left(\frac{V_p-v_g}{v_g}\right)\right]|a|^2\frac{\Omega}{\omega}.$$

Let us present our estimates of the demodulator parameters for a nonlinear optical planar waveguide with the section of 2 μm by 1μm, $n(\omega)=1.6$, and the length L=2 cm [6]. Let us take the refractive index of the microwave wave equal to 1.5, the power of the pump wave with $\lambda=1.3\mu$ equal to 2 mW, the microwave frequency equal to 100 GHz, and the nonlinear susceptibility equal to 2×10^2pm/V [7, 8]. Let us also assume that the wave velocities are equal ($V_p=v_g$). For these conditions, the nonlinear length L_{NL} and coherent length $L_c=\tau_c L_{NL}$ are equal to 9.9 cm and 3.1×10^2 cm, respectively. Since $L/L_c \ll 1$, we estimate the quantum efficiency by expression (17). The fraction in this expression is approximately equal to 1, since $|a|\approx 1$ according to (11) and $C_\pm \approx 1$ in the case of the weak dispersion. Therefore, the quantum efficiency is equal to $\eta \approx (L/L_{NL})^2 \approx 1/25$. The power conversion factor differs from this expression by the factor $\Omega/\omega \approx 4.3\times10^{-4}$.

The above estimate of the derivative $\partial\chi^{(2)}/\partial\omega$ is valid for the lack of spectral features in the vicinity of frequency ω. Otherwise, this estimate may differ from the real derivative by several times and even significantly more.

The attractive parametric amplification mode (see subsection 3.3) is possible only with a negative g. In turn, this is possible only with negative dispersion of the nonlinear susceptibility, according to expressions (18), (19).

Note that even in the case $L/L_c \ll 1$, corresponding to the linear regime, the quantum efficiency can be greater than 1 when $L/L_{NL}>1$ (see (17)). The reason for this is as follows. First, both the anti-Stokes quanta and quanta of pumping create the microwave photons (see Figure 2). Second, the number of quanta of pumping is much more than the number of the anti-Stokes quanta.

Acknowledgments
The study was supported by the Ministry of Education and Science of the Russian Federation (project 'II.10.2.1. Photonics of micro- and nanostructured media' state registration No. AAAA-A17-117060810014-9).

References

[1] Jager D 2009 *IEEE Microway Magazine* **10** 154

[2] Belkin M E and Sigov A S 2009 *Radiotehnika i elektronika* **54** 901

[3] Seeds A J and Williams K J 2006 *J. Lightwave Technol.* **24** 4628–4641

[4] Xie X, Zhou Q, Li K, Shen Y, Li Q, Yang Z, Beling A and Campbell J C 2014 *Optica* **1** URL `http://dx.doi.org/10.1364/OPTICA.1.000429`

[5] Sun K and Beling A 2019 *Applied Sciences* **9** 623 – 638 URL `www.mdpi.com/journal/applsci`

[6] Atutov S N, Mikerin S L, Plekhanova A I, Simanchuk A E, Sorokin V A, Yakimansky A V, Smirnov N N and Valisheva N A 2018 *Optoelectronics, Instrumentation and Data Processing* **54** 39–45

[7] Gorkovenko A I, Plekhanov A I, Simanchuk A E, Yakimanskii A V, Smirnov N N, Solovskaya N A and Nosova G I 2014 *Optoelectronics, Instrumentation and Data Processing* **50** 116–122

[8] Mikerin S L, Plekhanov A I, Simanchuk A E, Yakimanskii A V, Martynenkov A A and Valisheva N A 2018 *Optoelectronics, Instrumentation and Data Processing* **54** 385—-389 ISSN 8756-6990

Controlling the type and intensity of low-frequency waves generated by laser plasma clots in a force tube of magnetized plasma

A G Berezutsky[1], V N Tishchenko[1], A A Chibranov[1] I B Miroshnichenko[1,2], Yu P Zakharov[1], I F Shaikhislamov[1,2]

[1]Institute of Laser Physics SB RAS, 15B, Ac. Lavrentyev ave, Novosibirsk 630090, Russia
[2]Novosibirsk State Technical University, 20 K. Marx ave, Novosibirsk 630073, Russia

E-mail: a.berezuckiy@yandex.ru

Abstract In this work, we study the influence of the parameters of a magnetized background plasma on the intensity of whistler waves generated by periodic laser plasma bunches in a magnetic field tube. It is shown that at 0.3 < L_{pi} > 0.4 Alfvén waves and whistlers are generated. In the region L_{pi}> 0.5, intense whistlers with an amplitude of $\delta B_{max} / B_0 \sim 0.24$ are generated.

1. Introduction

Numerous works have been devoted to the interaction of laser radiation with matter, which is associated with a wide field of application - material processing, aerospace applications, special tasks, communications, as well as the emergence of a number of new directions in physics, chemistry, biology, instrument making, medicine, military technology, and sciences about life, etc. Through research on the interaction of laser radiation with matter, back in the late 60s, it became possible to generate laser plasma, which is a tool for laboratory modeling of nonstationary processes in space plasma. Laser plasma can act as a source of generation of artificial wave disturbances in the near-Earth plasma, which is an extremely urgent problem. Its solution is necessary both for the development of the problems of active experiments aimed at obtaining knowledge about the structure of the ionosphere, and for applications, most of which are related to the exploration of outer space.

In ILP SB RAS [1-2] it was shown for the first time that periodic plasma bunches created by repetitively pulsed laser radiation at a certain ratio between the energy and the pulse repetition rate generate a single wave in the surrounding gas, for example, an infrasonic wave. The effect of combining shock waves created by individual optical breakdowns of a gas into a single wave was called the wave merging mechanism (WMM). The length of a single wave linearly depends on the number of bunches in the packet, which makes it possible to form weakly damped energy flows with relatively low energy consumption. The method allows you to control the type, frequency and intensity of the generated waves. В настоящей работе представлены экспериментальные результаты резонансного взаимодействия сгустков лазерной плазмы с замагниченной плазмой фона для генерации альфвеновских и вистлерных волн.

This paper presents the experimental results of the resonant interaction of laser plasma bunches with a magnetized background plasma for the generation of Alfvén and whistler waves.

2. Wave merging mechanism

The wave merging mechanism (WMM) has been experimentally confirmed for gases, plasma with a magnetic field and under the action of a laser plasma source located in a magnetic field [3-8]. The essence of the mechanism is as follows. Periodic shock waves are created in a continuous medium, the initial velocity of which is greater than the speed of sound C_0. If the main criterion of the WMM is met, the waves from the individual bunches are combined and create a single low-frequency wave, the length of which is linearly dependent on the number of laser pulses.

The results of numerical simulations [4,6,9] showed that the MOV is effective for the formation of low-frequency Alfvén, magnetosonic and whistler waves in a magnetized plasma, the length of which linearly depends on the number of pulses and energy consumption for their creation. For single optical breakdowns, the wavelength weakly depends on the pulse energy (~ $Q^{1/3}$).

The main criterion for the WMM is the dimensionless pulsation frequency of repetitively pulsed laser radiation for combining waves, lying in the range: $0.1 < \omega = \frac{f \cdot R_d}{C_A} < 0.4$, where C_A is the Alfven velocity; Rd is dynamic radius (radius at which the pressure of the bunch is comparable to the pressure of the surrounding gas) $R_d = (Q/p_o)^{1/3}$, Q is the energy of the bunch; *f* is the laser pulse repetition rate. At frequencies ω <0.1, the bunches create a sequence of short noninteracting Alfvén waves. At large values of ω> 0.4, the wavelength decreases.

High efficiency of conversion of the source energy into a wave is achievable if additional conditions are met, depending on the properties of the surrounding background:

$$\alpha = \frac{R_L}{L_{pi}} > 5,$$

2. The Larmor radius is approximately equal to the radius of deceleration by the magnetic field RL = rl / Rd ~ 1. If these conditions are violated, the AQW is not formed or its amplitude is extremely small and unstable.

3. Plasma β is a parameter characterizing the ratio of the thermal pressure of the background plasma to the pressure of the external magnetic field [9]:

$$\beta = \frac{8\pi k\left(\left(1+Z_0\right)n_0\right)T_0}{B_0^2} < 1,$$

4. The ratio of the ion - inertial length to the dynamic radius of the plasma bunch:

$$L_{pi} = \frac{l_{pi}\left[cm\right]}{R_d\left[cm\right]} = \frac{c/\omega_{pi}}{R_d} \sim 0.1 \div 0.3,$$

The results of numerical modeling and experimental data [4, 6-9] showed that when the criteria of the WMM and the value of the dimensionless parameter (L_{pi}) <0.3 are met, alfven and magnetosonic waves are generated with a maximum amplitude at L_{pi} ~ 0.1, and when L_{pi}> 0.4, whistler waves are generated in the plasma. Intensity of which is characterized by the azimuthal component of the magnetic field Bφ.

The purpose of this work is to experimentally verify the generation of whistlers by laser plasma bunches. By varying the range of the dimensionless parameter L_{pi}, it is supposed to check the modes of wave generation: 0.1 <L_{pi} <0.3 bunches generate Alfvén and magnetosonic waves, and at 0.3 <L_{pi}> 0.4 - Alfvén waves and whistlers. Whistlers are generated in the region L_{pi}> 0.4.

3. Experimental setup

The experiments were carried out at the KI-1 facility of the Institute of Laser Physics of the Siberian Branch of the Russian Academy of Sciences. The experimental scheme is shown in Figure 1. A high-vacuum ($p = 10^{-6}$ Torr) chamber with a diameter of 1.2 m and a length of 5 m was filled with plasma generated by a θ-pinch, the concentration varied in the range $n_0 = 10^{12} \div 4 \cdot 10^{13}$ cm^{-3}. The plasma temperature, determined from the current-voltage characteristics of the probes, is approximately T ~ 10 eV. In experiments with an argon background Ar +, the typical velocity of plasma motion along the chamber axis is 10–20 km / s. A laser plasma bunches was generated by irradiating a polyethylene target with CO_2 laser pulses (spot area $S = 2.5$ cm^2). A target in the form of a plate, the plane of which is perpendicular to the magnetic field, is located on the axis of the camera. Pulses with an energy of Q ~ 200 J and a duration of ~ 1 μs sequentially created on the target bunches of laser plasma with an energy of Q ~ 25 J, an initial expansion velocity of 100 km / s. From measurements of the shape of the plasma glow at different times, it follows that the dynamics of the bunch expansion is close to spherical, which is necessary from the conditions for the manifestation of the WMM and the generation of quasi-stationary Alfvén and other types of waves localized in the flux tube. An external magnetic field is $B_0 = 100 \div 300$ G, along the chamber axis, is created by a source that supplies current to the solenoid, covering the entire outer surface of the chamber.

An array of eight sensors (Langmuir probes, magnetic probes and Rogowski belt) made it possible to measure the plasma concentration n, the components of the magnetic fields Bφ, Br, Bz, the electric field and the longitudinal current Jz in various sections of the chamber at a distance of z ~ 3 m from the target and r ~ 30 cm radially from the camera axis. The spatial resolution of the measurements is 1 cm, the temporal resolution is approximately 10 ns, the frequency band of the magnetic probes is 20 MHz.

The characteristic spatial scale of the problem is the dynamic radius R_d, which is the distance from the target at which the energy density of the laser plasma is compared with the background energy density. The main dimensionless parameters of the problem are the Alfven-Mach number (M_A) and the dimensionless parameter L_{pi} is the ratio of the ion-inertial length to the dynamic radius R_d and the magnetization of the ions R_L is the ratio of the gyroradius of the laser plasma ions to the dynamic radius R_d. Thermal beta background in these experiments was relatively low $\beta < 1$. To obtain data on the effect of the ion-inertial length, in our experiments, helium and ten times heavier gas, argon, were used as the background plasma.

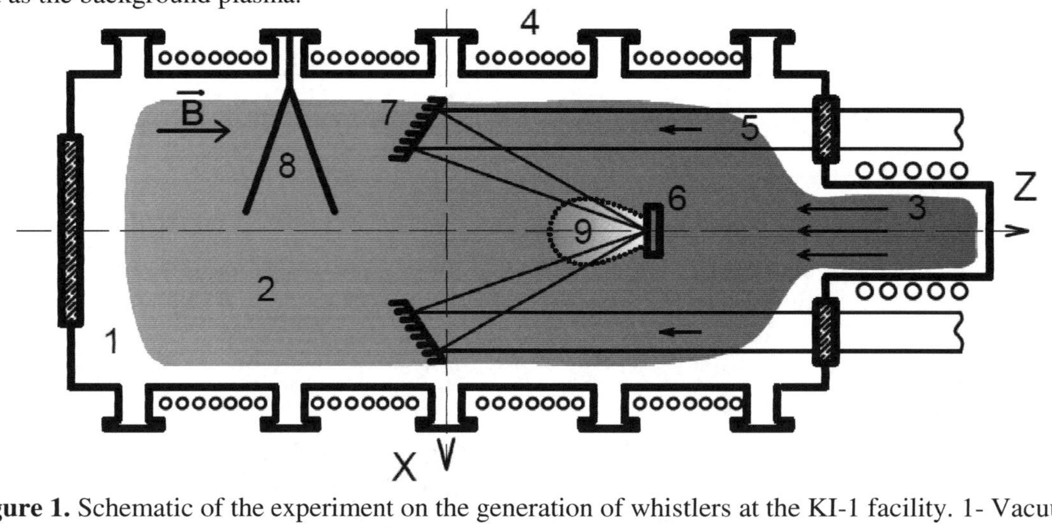

Figure 1. Schematic of the experiment on the generation of whistlers at the KI-1 facility. 1- Vacuum chamber. 2- plasma flux created by the θ -pinch (3). 4- turns of the solenoid, which creates an external magnetic field, axial to the axis of the chamber. 5- CO_2 laser radiation. 6- target. 7- focusing lenses. 8- measuring probes located in different sections of the chamber. 9- laser plasma bunches.

Table 1 shows the dimensionless parameters of the experiment.

Table 1. Experimental parameters

	Helium	Argon
B_0	100 G	100 G
R_d	~70 cm	70÷80 cm
M_A	1÷3,5	3,5÷10
$L_{pi}=c/R_d \cdot \omega_{pi}$	0,3÷0,6	0,6÷1,7
β	0,08÷0,8	0,08÷0,8
$\varepsilon_b=R_L/R_d$	~0,5	~0,4÷0,5

In terms of dimensional experimental parameters, the optimal conditions approximately correspond to the magnetic field B_0 ~ 100-200 G, the background plasma from argon ions at a temperature of ~ 10 eV, and the energy of laser bunches Q = 25 J at an initial expansion velocity of 150 km / s. The main difference between these experiments is the use of a background plasma of Ar and He, which, in comparison with experiments where a hydrogen plasma was used [7, 10], makes it possible to significantly increase the ion-inertial length.

4. Results

Figure 2 shows typical plasma concentration computed from measurements with Langmuir probes. Zero instant of time corresponds to the arrival of the laser pulse on the target. Up to this point, the signal indicates the background plasma concentration. The figure illustrates two cases - background plasma from He and Ar. The average velocity of the background plasma, measured by the time-of-flight method, was approximately V_{He} = 24 km/s and V_{Ar} = 12 km/s. To achieve lower background plasma concentrations $n_0 = (1 \div 2) \cdot 10^{12}$ cm^{-3} and, accordingly, a higher L_{pi} value, laser plasma was created at the stage of lowering the background plasma flux concentration, as can be seen from the figures.

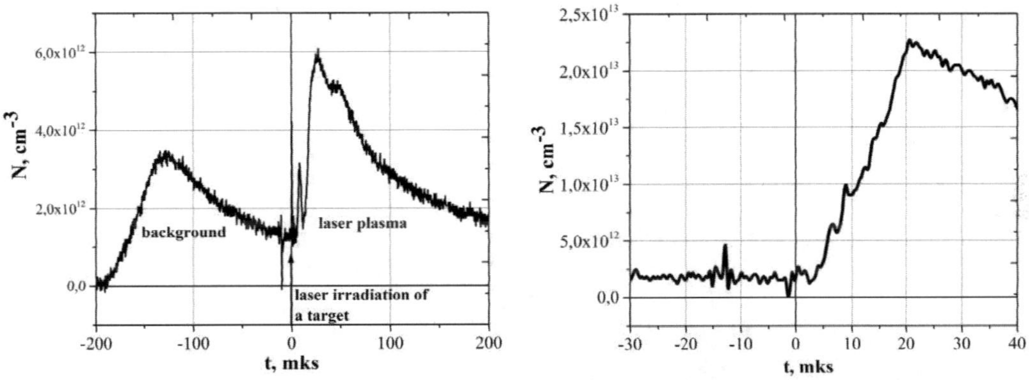

Figure 2. Plasma concentration n_0 (t) (helium on the left, argon on the right). The target is irradiated with a laser pulse at a background plasma concentration of n ~ $1.6 \cdot 10^{12}$ cm^{-3}. External magnetic field B_0 = 100 G.

An important result of the experiments was the registration of the characteristic properties of whistlers: an azimuthal magnetic disturbance of right-handed circular polarization propagating at high speeds (Vw> 300 km / s). Figure 3 shows typical oscillograms of the azimuthal component of the magnetic field $B\varphi$ (t). Registration of signals by magnetic probes at distances z = 140 cm showed that magnetic disturbances have right-hand circular polarization and a characteristic decreasing frequency from 10^6 Hz at the beginning of the packet to $2 \cdot 10^5$ Hz at the end.

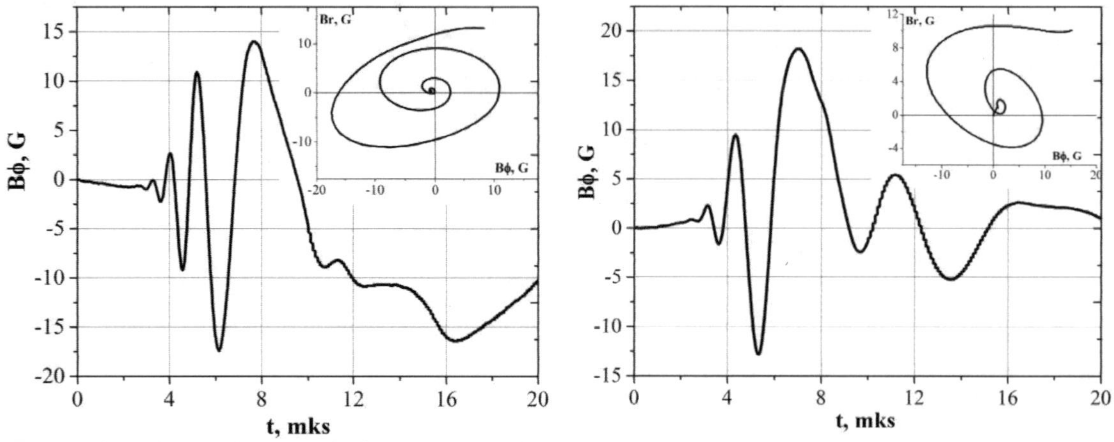

Figure 3. Azimuthal magnetic field Bφ (t) (left - background helium plasma (L_{pi} = 0.45), right - argon plasma (L_{pi} = 0.6)). The inset shows the hodograph of the magnetic field. Distance z = 140 cm from the target. External magnetic field B_0 = 100 G.

To illustrate the transient regime, when the parameter of ion-inertial length lies in the range $0.3 < L_{pi} < 0.4$, Figure 4 shows the case when, in addition to whistlers, an Alfvén wave propagates in a medium [10]. The disturbance was registered at a distance z = 90 cm from the target. The inset shows the hodograph of the magnetic field. At a concentration of hydrogen plasma $n_0 = 5 \cdot 10^{12}$ cm^{-3}, a whistler propagates in the medium at a speed of Vw = 220 km / s, and an Alfvén wave (C_A = 70 km / s). The dot on the graph shows the arrival time of the Alfvén wave at the probe.

This is also confirmed by the hodograph of the magnetic field, which has right-handed circular polarization, which is typical for whistlers. At the moment of time t = 12 μs, the polarization is transformed into a left-handed one, which is associated with the arrival of an Alfvén wave.

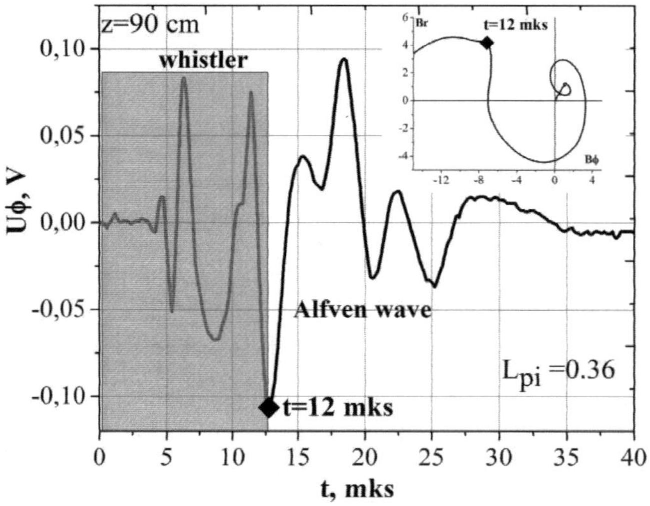

Figure 4. Derivative of the azimuthal magnetic field generated by one plasma bunch. External magnetic B0 = 175 G. The inset shows the hodograph of the magnetic field. The dot on the graphs shows the arrival time (t = 12 μs) of the Alfvén wave.

Thus, when the parameter L_{pi} lies in the range $0.3 < L_{pi} < 0.4$, the regime of generation of whistler and Alfvén waves is observed in a plasma with a magnetic field. The amplitude of the whistlers $B_{\varphi w}$ =

8 G, the Alfvén wave $B_{\varphi a}$ = 12 G. In this case, the amplitude of the waves can be increased when switching from the mode of generating several types of waves to the mode of generating only whistlers or Alfvén and magnetosonic waves.

5. Summary

In this work, we investigated the influence of the parameters of the magnetized background plasma on the intensity of whistler waves generated by periodic bunches of laser plasma in the magnetized background plasma. Experimentally, at the KI-1 facility, the range of values of the dimensionless parameter L_{pi} was confirmed, at which three modes of wave generation are observed. In the range 0.1 <L_{pi} <0.3, bunches generate Alfvén and magnetosonic waves [8], and at 0.3 <L_{pi}> 0.4 - Alfvén and whistler waves. In the region L_{pi}> 0.5, intense whistlers with $\delta B_{max} / B_0 = 0.24$.

Acknowledgements

This work was supported by ILP program № 0243-2021-0003.

References

[1] Tishchenko V N, Apollonov V V, Grachev G N, Gulidov A I, Zapryagaev V I, Men'shikov Ya G, Smirnov A L and Sobolev A V 2004 *Quantum Electron.* **34** 10 pp 941–947
[2] Grachev G N, Ponomarenko A G, Tishchenko V N, Smirnov A L, Trashkeev S I, Statsenko P A, Zimin M I, Myakushina A A, Zapryagaev V I, Gulidov A I, Boiko V M, Pavlov A A and Sobolev A V 2006 *Quantum Electron.* 36:5 pp 470–472
[3] Tishchenko V N, Posukh V G, Gulidov A I, Zapryagaev V I, Pavlov A A, Boyarintsev È L, Golubev M P, Kavun I N, Melekhov A V, Golobokova L S, Miroshnichenko I B, Pavlov A A and Shmakov A S 2011 *Quantum Electron.* **41** 10 pp 895–900
[4] Tishchenko V N, Shaikhislamov I F and Berezutskiy A G 2014 Supercomputers Technologies in Science, Education and Industry: The almanac Moscow, MSU Publ. pp 65–74 (in Russian)
[5] Tishchenko V N and Shaikhislamov I F 2010 *Quantum Electron.* **40** 5 pp 464–469
[6] Tishchenko V N and Shaikhislamov I F *Quantum Electron. 2014* **44** 2 pp 98–101
[7] Tishchenko V N, Zakharov Yu P, Shaikhislamov I F, Berezutskii A G, Boyarintsev E L, Melekhov A V, Ponomarenko A G, Posukh V G and Prokopov P A 2016 *JETP Letters.* **104** 5 pp 293–296
[8] Tishchenko V N, Berezutskii A G, Boyarintsev È L, Zakharov Yu P, Melekhov A V, Miroshnichenko I B, Ponomarenko A G, Posukh V G and Shaikhislamov I F 2017 *Quantum Electron.* **47** 9 pp 849–852
[9] Berezutskii A G, Tishchenko V N, Zakharov Yu P, Miroshnichenko I B and Shaikhislamov I F 2019 *Quantum Electron.* **49** 2 pp 178–180
[10] Prokopov P A, Zakharov Yu P, Tishchenko V N, Boyarintsev E L, Melekhov A V, Ponomarenko A G, Posukh V G and Shaikhislamov I F 2016 *Solar–Terrestrial Physics.* **2** 1 pp 19–33

Laboratory modeling of a field-aligned currents system generated by a flow of inner magnetospheric plasma

A Chibranov, A Berezutsky, M Efimov, Y Zakharov, I Miroshnichenko, V Posukh, M Rumenskikh, P Trushin, I Shaikhislamov

Institute of Laser Physics SB RAS, pr. Lavrentyeva 15B, Novosibirsk, 630090, Russia

E-mail: chibranov2013@yandex.ru

Abstract. For the first time in laboratory conditions, an experiment to simulate a system of field-aligned currents arising on planets such as Hot Jupiters in the presence of dense inner-magnetospheric plasma was carried out. The magnitude and transit time of field-aligned currents were measured as a function of the magnetic field using flat electrodes. The geometry of the expansion of plasma streams was pictured by gated camera. Also, in a first approximation, the efficiency of energy transfer from plasma flows to field-aligned currents was calculated. The results obtained create a basis for future laboratory experiments on this topic and improve existing numerical models.

1. Introduction

Despite the great interest shown today in the study of space, many of its processes still remain difficult for studying. One of such processes is a formation of global field-aligned currents in presence of dense plasma flows in the planetary magnetospheres. This phenomenon is typical of Jupiter, and has recently gained relevance due to the discovery of exoplanets such as Hot Jupiters. The magnetospheres of hot exoplanets are characterized by ionization and outflow of the upper atmosphere.

Field-aligned currents transfer energy from the boundary layer and the magnetodisk to the polar ionosphere, causing various effects. The presence of planetary plasma or planetary wind, formed by the upper atmosphere ionization and held by the planet's magnetic field, with a plasma density several orders of magnitude greater than the external stellar plasma density, can lead to new phenomena that at the moment can only be detected in a laboratory experiment, because numerical modeling of such a system remains difficult due to the complex topology.

This paper presents the results of experiment on modeling of strong field-aligned currents, which are generated inside the magnetosphere [1] and make a closure at the planet poles. The experiment was carried out at the Laboratory of High Power Lasers of ILP SB RAS using the KI-1 facility [2-3]. A distinctive feature of the experiments is the use of energetic plasma flows generated by a powerful pulsed CO_2 laser. In previous laboratory experiments on laser plasma flow around a magnetic dipole [4-8], it was shown that, in addition to the formation of a clearly defined frontal region of the magnetosphere, the external plasma flow generates at the poles an intense system of field-aligned currents. Detailed measurements of the total magnitude and local density of field-aligned currents, magnetic disturbances at the poles revealed their similarity to the currents of Region-1 in the Earth magnetosphere. Laboratory field-aligned currents were observed only if they could be closed through the conducting surface of the dipole. The novelty of these experiments lies in the fact that field-aligned currents are formed this time by an inner-magnetospheric plasma flow.

2. Experimental setup

A dipole magnetic field source (Fig. 1) was placed inside a large cylindrical vacuum chamber of 5 ×1.2 m size with a residual pressure of to 10^{-6} Torr. The source consisted of two solenoids connected serially each of which could create a magnetic moment of up to 10^6 G·cm^3 powered by pulsed ~ 200 μs voltage generator. The magnetic moments of the coils were directed parallel to each other, creating a common dipole field. This dipole configuration made it possible to obtain magnetic fields with a large moment at a lower applied voltage. Polyethylene (C_2H_4) targets were fixed on both sides of the dipoles housing at the magnetic poles. Targets were symmetrically and simultaneously irradiated with two 100 ns pulses of a CO_2 laser with an energy of E = 200 J, creating plasma flows with a concentration of up to n = 10^{14} cm^3, velocity υ = 100 km/s and total kinetic energy E = 25 J. The plasma ablated from the target simulated an inner-magnetospheric plasma of exoplanets called hot Jupiter.

To measure the electric potential and field-aligned currents, flat copper electrodes were used, fixed in pairs on the target at each dipole pole (U1 and U2 in Fig. 1b). One pair of electrodes was electrically gaped to measure the potential, and the second was short-circuited with a copper bus, which passed through the Rogowski coil to measure the current I. To obtain plasma images, high speed four gated camera HSFC PRO was used.

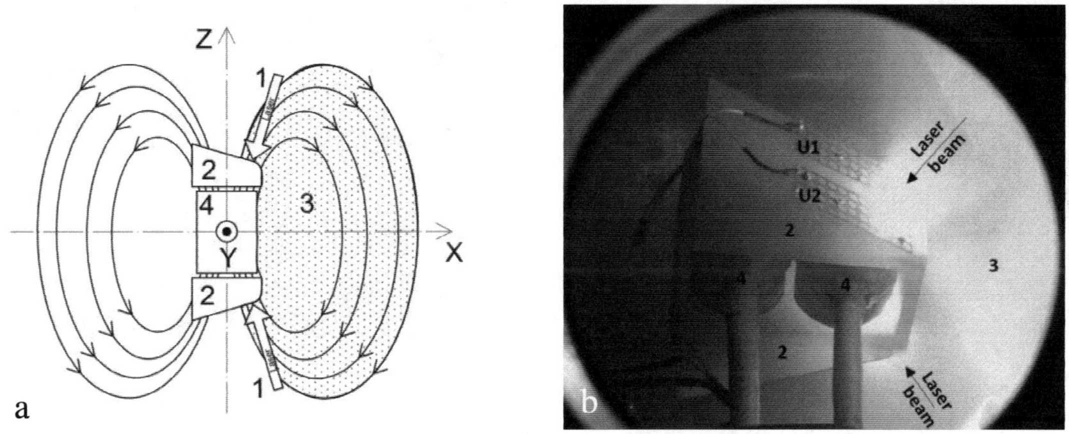

Figure 1. Scheme (a) and photo (b) of targets-dipole system with a generated plasma. 1 - laser beams, 2 - polyethylene targets, 3 - plasma flow, 4 - dipole magnetic field source, U1 и U2 – copper measuring electrodes.

3. Results

The field-aligned currents arising during the expansion of plasma flows inside the dipole magnetic field with the moment M = 10^6 G·cm^3 reached a magnitude of 1.77 kA (Fig. 2a). The value of such currents directly depended on the magnetic moment of the dipole field (Fig. 3 - black curve). The transpolar potential reached 275 V. For comparison, measurements were carried out without dipole field producing a relatively small current I = 77 A between the electrodes (Fig. 2c), apparently caused by a photo-current because of uneven irradiation of copper electrodes with laser radiation. The images demonstrate that in the presence of the dipole field the plasma expands along the field lines (Fig. 2b), while without magnetic field the quasi-spherical plasma clouds are formed (Fig. 2d).

It was found that the time duration of field-aligned linearly depends on the dipole magnetic moment (Fig. 3 - red curve). For example, at a magnetic moment M = 0.23 · 10^6 G·cm^3, the current existed 2.5 μs, and at M = 10^6 Gs · cm^3, it was 3.9 μs. This may be due to the fact that a magnetic field with a larger moment captures the plasma flow for a longer time, allowing it to form field-aligned currents for longer duration.

Journal of Physics: Conference Series **2067** (2021) 012020 doi:10.1088/1742-6596/2067/1/012020

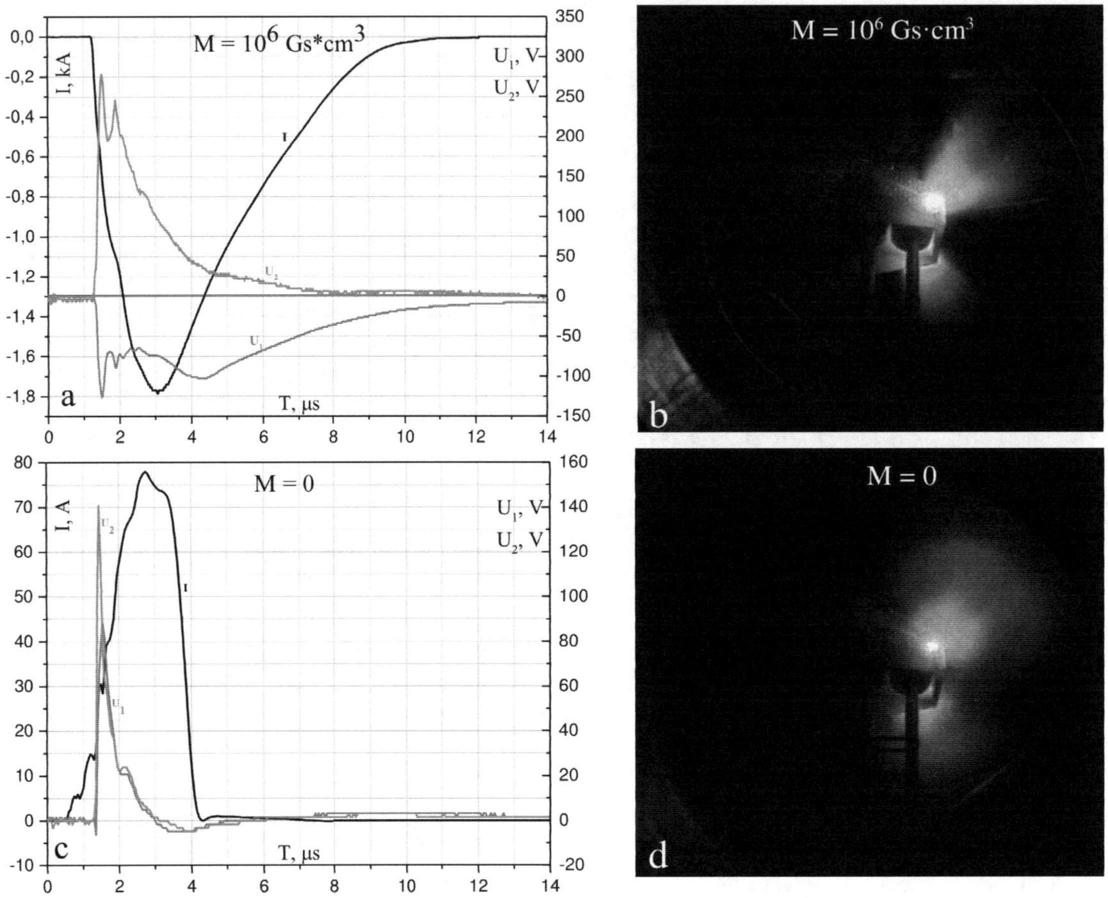

Figure 2. a and c - The current measured by the Rogowski coil (left axis) and the electric potential from two electrodes (right axis) during the expansion of laser plasma inside the dipole field with a magnetic moment of M = 10^6 Gs · cm^3 and without magnetic field, respectively. c and d – plasma images obtained at a time of 1.8 μs after irradiation of the targets.

Also, as a first approximation, it is possible to determine the total energy transferred from the plasma flow to the field-aligned currents. The energy of the plasma flow can be obtained as

$$E_p = 4\pi R^2 \int 0.5\, n_i\, m_i (R/t)^3 dt \approx 25\, J,$$

where n_i – plasma density measured by Langmuir probe at a distance *R* from the focal spot, m_i - ion mass. The measured value of the transpolar potential and total current allows to determine the electric energy of the field-aligned currents

$$E_c = \int IU\, dt \approx 1.1\, J,$$

where *I* - total current, *U* - transpolar potential. That is, the energy transfer efficiency from plasma flows to field-aligned currents was about 4.4%.

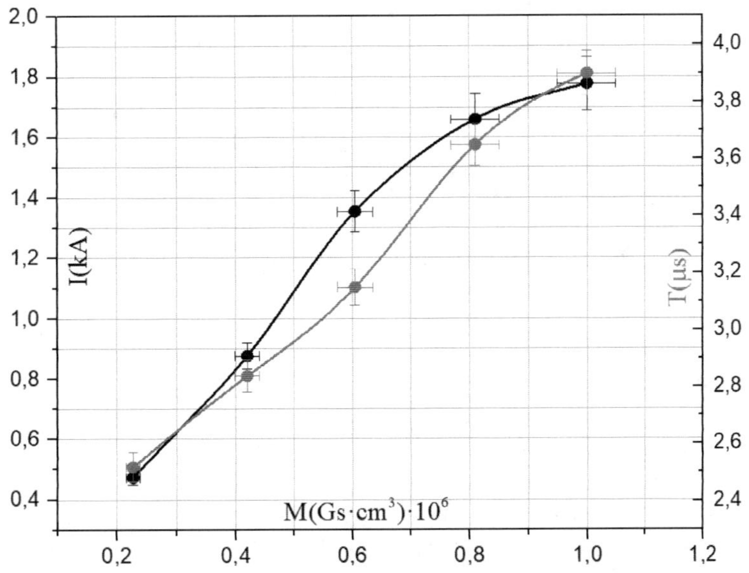

Figure 3. The amplitude (black - left axis) and duration (red - right axis) of the field-aligned current as a function of the magnetic moment of the dipole field.

4. Conclusion

In recent years, planets such as hot Jupiters have been actively discovered in large numbers, which can have inner-magnetospheric plasma flows inside the intrinsic magnetic field. On such planets, various new physical phenomena can be expected, one of which is field-aligned currents caused by the motion of plasma along the expanding lines of force of the magnetic field. Field-aligned currents are an important element of planetary magnetospheres, transferring energy from the faraway regions and magnetodisk to the polar ionosphere, causing such a key effect as the generation of radio emission.

Currently, the study of planets such as hot Jupiters is mainly carried out using numerical modeling, the input data for which are obtained using spectral observations of such planets. For a full-fledged numerical simulation of complex processes occurring on hot Jupiters this information is insufficient, and laboratory simulations can be used to provide additional data and physical insight.

This work was the first to demonstrate the possibility of laboratory modeling of field-aligned currents arising on planets such as hot Jupiters. To create inner-magnetospheric plasma flows, irradiation of polyethylene targets with a powerful laser pulse was used. During the plasma flows expansion in a dipole field with a magnetic moment of up to $M = 10^6$ G·cm^3, field-aligned currents were measured reaching I = 1.77 kA. The amplitude and duration of field-aligned currents directly depended on the magnetic field. The results obtained in this work create the basis for future laboratory experiments on this topic, which will improve the models of planets such as Hot Jupiters.

Acknowledgments

The work was supported by RFBR grant № 18-2921018, RNF grant 18-12-00080 and the grant of the Ministry of Science and Higher Education of the Russian Federation (No 075-15-2020-780).

References

[1] Efimov M A *et al* 2021 Capture of a dipolar magnetic field by laser-produced plasma *Quantum Electron.* ***51*** 3 p 222

[2] Gevorkyan L B, Orishich A M and Ponomarenko A G 1976 On the development of methods for modeling some astrophysical phenomena in a laboratory experiment *Collection of scientific papers Aerophysical research* **6** pp 192-196 (in Russian)

[3] Zakharov Yu P, Orishich A M and Ponomarenko A G 1988 Laser plasma and modeling of nonstationary space processes *Collection of scientific papers Aerophysical research p 220* (in Russian)

[4] Shaikhislamov I F, Antonov V M, Zakharov Y P, Boyarintsev E L, Melekhov A V, Posukh V G and Ponomarenko A G Laboratory simulation of field aligned currents in an experiment on laser-produced plasma interacting with a magnetic dipole *Plasma Physics and Controlled Fusion* ***51*** 10 105005

[5] Ponomarenko A G, Zakharov Yu P, Antonov V M, Boyarintsev E I, Melekhov A V, Posukh V G, Shaikhislamov I F and Vchivkov K V 2008 Simulation of strong magnetospheric disturbances in laser-produced plasma experiments *Plasma Physics and Controlled Fusion* ***50*** 7 074015

[6] Antonov V M, Boyarintsev E L, Zakharov Y P, Melekhov A V, Posukh V G, Ponomarenko A G and Shaikhislamov I F 2010 Effect of Surface Conductivity on Magnetosphere Formation in Experiments with Laser Plasma flow over a Magnetic Dipole *Journal of applied mechanics and technical physics* ***51*** 5 pp 340-348

[7] Shaikhislamov I F, Antonov V M, Boyarintsev E L, Posukh V G, Melekhov A V, Zakharov Y P and Ponomarenko A G 2012 Measurement of the transpolar potential in laboratory magnetosphere *Cosmic Research* ***50*** 6 pp 410-420

[8] Shaikhislamov I F, Zakharov Y P, Posukh V G, Boyarintsev E L, Melekhov A V, Antonov V M and Ponomarenko A G 2011 Laboratory experiment on region-1 field-aligned current and its origin in the low-latitude boundary layer *Plasma Physics and Controlled Fusion* ***53*** 3 035017

Journal of Physics: Conference Series **2067** (2021) 012021 doi:10.1088/1742-6596/2067/1/012021

On the opportunity of Laser Plasma simulation of Plasma Jets formation in moderate magnetic fields ~ kGs

Yu P Zakharov[1], V P Neznamov[2], V A Terekhin[2], I F Shaikhislamov[1], V G Posukh[1], P A Trushin[1], A A Chibranov[1], A G Berezutsky[1], M S Rumenskikh[1], M A Efimov[1]

[1]Institute of Laser Physics (ILP), Siberian Branch of Russian Academy of Sciences (SB RAS), Novosibirsk, Av. Lavrentyeva 15Б, 630090, Russia.
[2]All-Russian Research Institute of Experimental Physics (VNIIEF), Sarov, Av. Mira 37, 607188, Russia.

E-mail: ki1z@mail.ru

Abstract. Today a number of various experiments with Laser-Produced Plasmas (LPP), done at very-high Magnetic Fields (up to $B_0 \geq$ MG) are using to model the physics of Astrophysical and Space Jets, a various processes of their formation and possible long-range propagation at various angles ϑ to magnetic fields. We discuss the opportunity and present the first results of new-type experiments on simulation of Jets with LPP at KI-1 facility of ILP, at moderate magnetic field ~ kGs, oriented quasi-transverse ($\vartheta \approx 60^0$) of LPP-blob expansion (with velocity $\mathbf{V}_0$) relative to $\mathbf{B}_0$. They were done on the base of all our preliminary studies, both at large-scale, high-vacuum chamber (∅120 cm of KI-1) and others devices with LPP (oriented earlier at V_0 transverse to B_0, with $\vartheta = 90^0$).

1. Introduction

Since 1991 a lot of investigations at **KI-1** facility of ILP were done [1-7] on the expansion of LPP across magnetic field B0, including a few cases with multi-kGs ones, of dipole type [1-3, 5] and quasi-uniform [4, 6, 7] of small Helmholtz Coils (∅ ≤ 15 cm). Therefore, a data about revealed Jet-like structures (in both cases) were very preliminary, and only recently we could study initial stage of Jet formation from LPP (with the flat target) in the field B_0 = 300 Gs [8] at scale ~ 100 cm, comparable with the diameter ∅ 120 cm of vacuum chamber **KI-1**. Now, with the purpose to develop the methods of laboratory simulation in the LPP-experiments, into the center of chamber was installed a spherical plastic target ∅ = 10 mm with the system of its symmetrical irradiation by 4 beams (of tetrahedral scheme). As a result, we had started a new kind of experiments [**10**] to generate a spherically-symmetrical plasma cloud and to study its collisionl**ess** interaction with magnetic field and magnetized plasma background (for verification of new 3D-Codes). In our previous experiment **MHD** [7] with a small spherical target (∅ = 3 mm and 2-beams irradiation), in the field 8 kGs we had observed formation of 2 Jets (along to beams) with velocity $V_J \approx 50$ km/s ~ $V_0/3$ (at radius R > 2 cm). For the initial total energy of this LPP, E_0 ~ 10 J, its deceleration radius [11] by field is $R_B = (3E_0/B_0^2)^{1/3} \approx 1{,}7$ cm and directed ($\propto V_0$) ion Larmor radius – $R_L \approx 0{,}5$ cm, so, the criterion of ion magnetization had enough small value $\varepsilon_b = R_L/R_B \approx 0{,}3 < 1$ for the effective interaction of LPP with B_0–field at scale ~ R_B, including its deceleration and cavity formation [7].

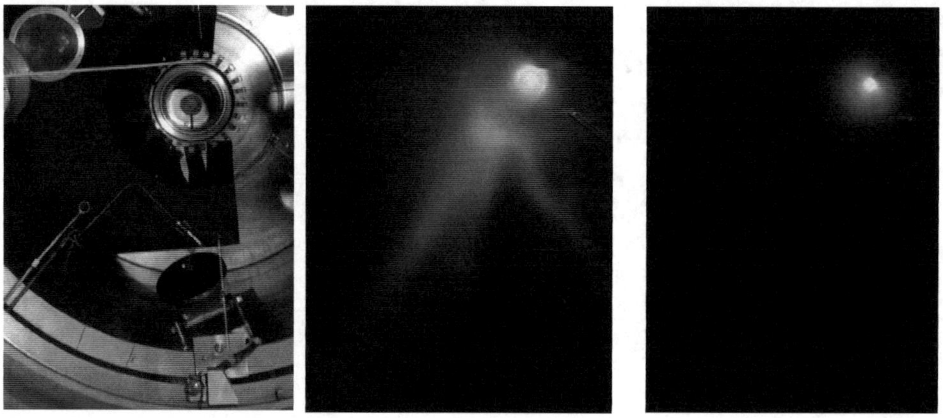

a **b** **c**

Figure 1. Time-integrated photos (in almost the same scales) of interior, inside of **KI-1** vacuum chamber: **a**) part of optical scheme with only **2** used (here) reflecting copper mirrors – at the bottom and left (fixed on two rings ∅ 110 cm, from stainless steel); **b**) LPP **red Jet** in magnetic field B_0 = 400 Gs (with H_2-gas, **P** = 1,5 mTorr); **c**) with gas, but without magnetic field. At **a**) the pair of magnetic probes are seen in the middle between mirrors. The first **IAP**-probe (from the center) has ∅**1,75** cm and second **RM**-probe is suited at distance L ≈ 6 cm from the first one.

2. Experiments on generation and study of Jets in moderate magnetic fields 300÷400 Gs

Therefore we had tryed to realize in the given 4-beams scheme, attempt to form a Jet-like structure from a larger spherical target (C_2H_4), by using only 2 beams, from the left entrance laser window). At Figure **1** one can see the relevant part of this scheme with 2 copper mirrors (at the bottom and at left), reflected 2 beams (here) of CO_2 -laser to the target above them. As a result, with magnetic field B_0 = 400 Gs, we could registered by time-integrated photos (Figure **1a**), a long red plasma stream from target downwards at presence of H_2-gas (with pressure ≈ 1,5 mTorr). With gas, but without field, only small blue sphere appears shifted from target (Figure **1c**). The photo (Figure **2**) of Gated Optical Imager (**GOI**,) in the same field (in vacuum) have revealed (at time t = 2,5 μs with exposition 1,5 μs, without gas) a Jet-like structure (of size up to S≈ 25 cm into the same down-left direction, with the velocity V_0 ≈ 100 km/s, rather well corresponding to the initial expansion velocity of LPP- sphere, generated in the case of total 4 beams (when LPP energy E_0≈25 J).

The red color of observed luminosity well corresponds to the Balmer Alfa (656.2 nm) of Hydrogen and we also seen such an effect before, during previous experiment **Super-Jet** [8] with the flat target (C_2H_4) and real kinetic energy of LPP 50 J (with effective energy 300 J, into 4π) in the field B_0 = 300 Gs. In the given case, with expected LPP kinetic energy E ≈ 12 J and B_0 = 400 Gs, expected cavity radius could be $R_B = (3E/B_0^2)^{1/3}$ ≈ **13** cm and ion Larmor R_L≈ **10** cm, and so the ε_b ~ 1 and we have presented at Figure **2**, a **GOI**-data on Jet-like structure up to the radius R ≈ 25 cm (see Figure **2**.).

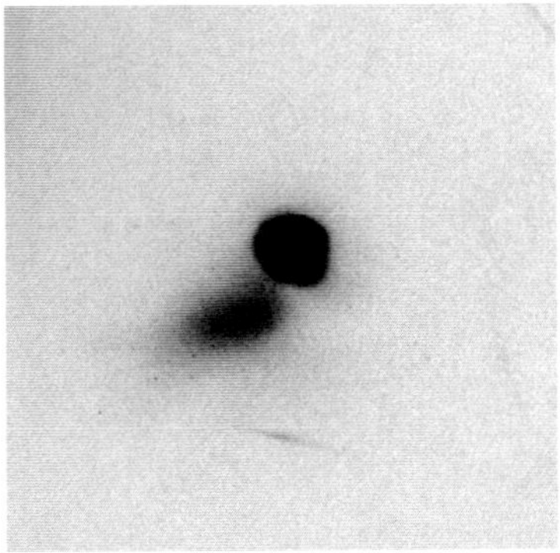

Figure 2. **GOI**-data on the **Jet**-like plasma expansion (from **2** laser beams) at the moment t=**2** мкс, with exposition τ = **1,5** μs, (negative) in vacuum 3*10^{-6} Torr at field B_0=400 Gs. The dark spot in center is a slow plasma near target, while below – fast stream with a front velocity V_0= R/t ≈125 km/s close to the initial velocity of LPP expansion without field. This velocity was measured by Langmuir probe (at its maximal R = 25 см), which support are seen lower.

3. Magnetic probes data on propagation "2-Beams" Jets at large distances with B_0 = 400 Gs

Since it was impossible to study such Jets at distances R > 25 см by GOI (to get such **GOI**-Photos in later times, we need to increase **P** more than 5÷7 mTorr) and also by Langmuir probes (in this direction), we did additional measurements by the pair of magnetic probes (see Figure **1a**, at the left). Their main data on magnetic disturbances by Jet are presented at Figures **3, 4, 5**.

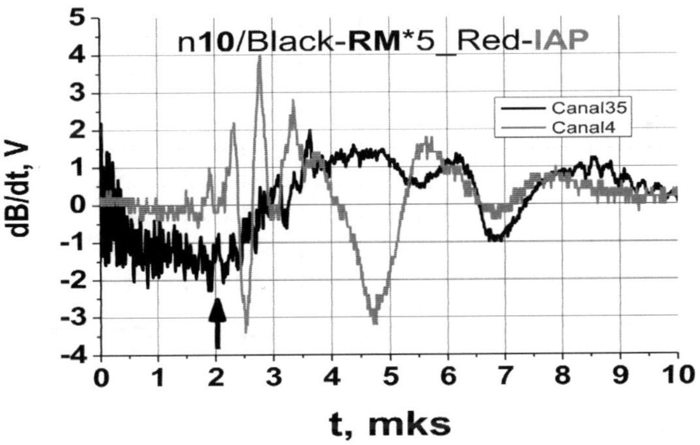

Figure 3. Example of preliminary data of magnetic probes signals obtained at distances $\mathbf{R_1}$ = **26** cm (**IAP**-probe) and $\mathbf{R_2}$ = **32** cm (**RM**) by **black**, it is magnified by factor **5**). They are oriented mutually by orthogonally, with effective area $\mathbf{S_1}$ = **16,5 см²** and $\mathbf{S_2}$ = **2 см²**.
IAP – oriented to measure $\mathbf{B_\varphi}$, and
RM - to measure $\mathbf{B_z}$ variations.

Here (above) at the moment **t** =**2** mks (by black arrow at bottom) is marked the moment, when the decline of $\mathbf{B_z}$-derivations (**RM**-probe) was stopped, before its rising (data of 10 September 2021). Laser shot was done at the moment **t** = **0**.

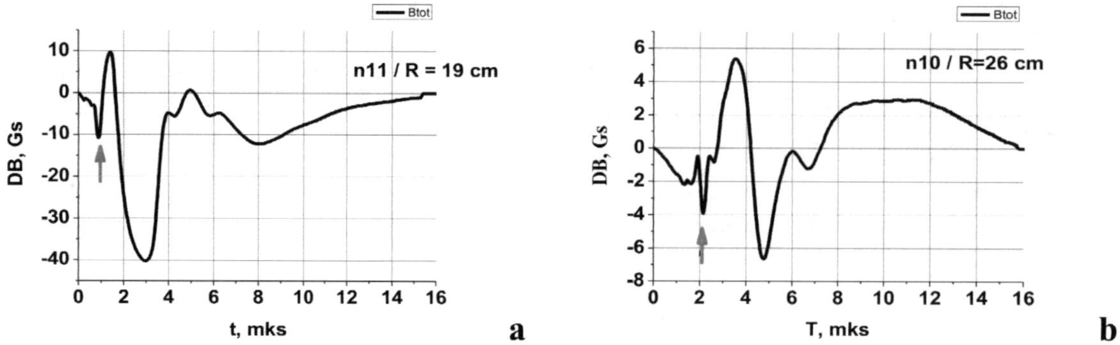

Figure 4. Examples of the moments (by red arrows) when a spikes of $\mathbf{B}_{\varphi}$-field were registered by **IAP**-probe in front of **Jets** arriving to probe.

At the Figure **5**, the main moments of **Jets** arriving to probe **RM** are presented, first of all, the **a**-time **2** mks (of **RM** at distance **25** cm), which is rather close to **AIP**-signal at distance 26 cm (at Figure **4b**). Therefore we could combine the both data into general **R-t** diagram at Figure **6**.

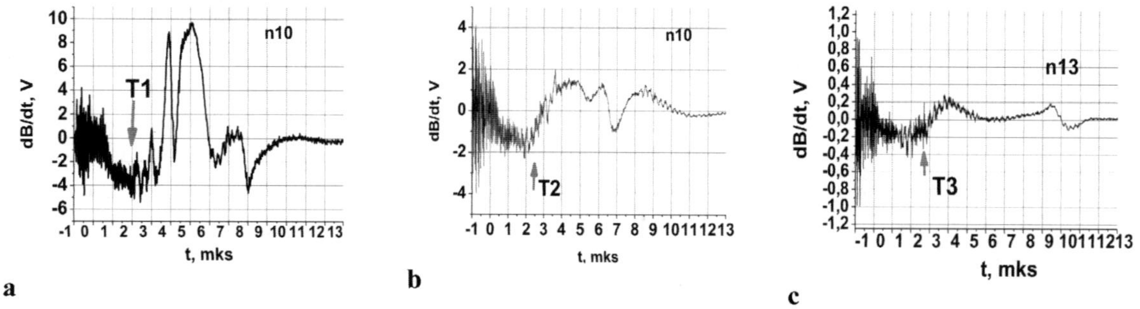

Figure 5. The data on the dynamics of magnetic disturbances Derivatives (**D**) at various distances of magnetic probe **RM** (**25**, **32** and **35** cm) at the moments **T1** = **2**, **T2** = **2,5** and T3 ≈ **2,8** mks (when the decline of **D** stops). Initial decline of **D** explains by magnetic field $\mathbf{B}_z$ exclusion near target and it stops, when the LPP reaches the probe. These moments are well correlated with the spikes in $\mathbf{B}_{\varphi}$–field (Figure **4**), registered by **IAP**-probe, as a result, a joint **R-t** diagram of Jets front was found (Figure **6**).

4. Discussion and Conclusions

Presented at Figure **6** (below), **R-t** diagram of Jet front, corresponds to its motion without deceleration accordingly insufficient level of ion-magnetization on $\varepsilon_b = R_L/R_B \approx 1$ [7]. Since in the given case ion-Larmor radius $R_L \approx 10$ cm and we register Jet propagation up to the length $L \approx 50 \div 60$ cm (see Figure **1a,b**), we can conclude that also in the case of low-level of ion-magnetization, the Jets could propagate in the Regime of Long-Range Propagation of Large-Scale Jets [12].

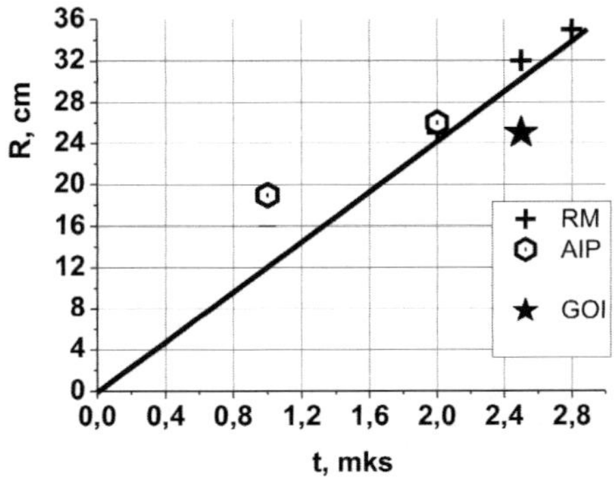

Figure 6. Joint **R-t** diagram of **Jet** front by the data: **RM** – probe, **IAP** –probe, **GOI** – Gated Optical Imager (photo at Figure **2**). The Line corresponds to the velocity 120 km/s, rather close to the initial velocity of LPP expansion (V_0 ~ 100 km/s) in the full 4-beams scheme [10], in contrast to our previous data [7], with multi-kGs field and with high-level of ion-magnetization on $\varepsilon_b \approx 0,3<1$, While in the given case we have $\varepsilon_b \approx 1$, only.

This work was done due to Financial Support of the Program n**5-5** of NCPM (VNIIEF) and in part by Program of ILP N0243-2021-0003.

References

[1] Zakharov Yu P, Dynamics and energetics of plasma blob expansion transverse to Dipole magnetic field at presence of background. *Report of RFBR* 95-02-04605a, 25 p. (1995).
[2] Zakharov Yu P, Melekhov A V, Posukh V G, et all, Dynamics of Laser-Produced Plasmas in Dipole magnetic field under conditions of near-Earth anti-asteroidal explosions. *Proceedings of the 8th International Conference on Plasma Physics (Nagoya, 9-13 Sept. 1996)* **2** 1678.
[3] Zakharov Yu P, Laboratory simulation of transverse magnetic field effects on dynamics on dynamics of Plasma Streams in Magnetosphere. In 5th International Symposium *"Double Layers/Potential Formation and Related Nonlinear Phenomena in Plasmas"* in "Plasma Forum" (Sendai, 17-19 Sept. 1996), "World Scientific" (1997) 225.
[4]. Zakharov Yu P, Eremin A V, Orishich A M, et al., Laboratory simulation of depolarization and deflection of Plasma Jets drifting across Magnetic Field in Geoplasma background. *XXIV Intern. Conf. on Phenom. In Ioniz. Gases*, **II** (Warsaw, 11-16 July (1999) 97.
[5] Zakharov Yu P 2002 *Advances Space Research* **29** 1335
[6] Zakharov Yu P, Eremin A V, Orishich A M, et all, 2002 *Advances in Space Research* **29** 1351
[7] Zakharov Yu P, Antonov V M, Boyatintsev E L, 2006 *Plasma Physics Rep*. **32** 183
[8] Zakharov Yu P, Ponomarenko A.G., Terekhin V A, Posukh V.G, Shaikhislamov I F, Chibranov A.A, 2019 *Quantum Electronics* **49** 181
[9] Zakharov Yu P, Neznamov V P, Terekhin V A, Shaikhislamov I F, Posukh V.G, et al 2021 On the opportunity of Laser Plasma simulation of Plasma Jets formation in moderate magnetic fields ~ kGs. *Modern Problems of Laser Physics - MPLP-2021.The IX International Symposium Technical Digest. Novosibirsk, 2021,* **173**. Submitted to Quantum Electronics
[10] Zakharov Yu P, Terekhin V.A., Shaikhislamov I.F., et al. 2021 Spherical Clouds of Laser-Produced Plasma to model 3D-dynamics of Artificial Plasma Releases in near-Earth space. *Modern problems of laser physics - MPLP-2021. The IX International Symposium technical Digest. Novosibirsk*, 170
[11] Raizer Yu P, 1963 *Zhurnal Prikl. Mekhan. Tekh. Fiz.* N6 19 – In Russian.
[12] Dmitriev A V, Suvorova A V 2015 Large-Scale Jets in the magneto sheath and plasma penetration across the magnetopause: THEMIS observations *J. Geophys. Res.*, **120A**, 4423.

AUTHOR INDEX